Chemisorption
and Magnetization

Chemisorption and Magnetization

P. W. Selwood

Department of Chemistry
University of California
Santa Barbara, California

ACADEMIC PRESS   New York   San Francisco   London 1975

A Subsidiary of Harcourt Brace Jovanovich, Publishers

ACADEMIC PRESS, INC.
111 Fifth Avenue, New York, New York 10003

United Kingdom Edition published by
ACADEMIC PRESS, INC. (LONDON) LTD.
24/28 Oval Road, London NW1

Library of Congress Cataloging in Publication Data

Selwood, Pierce Wilson,   (date)
     Chemisorption and magnetization.

     Edition for 1962 published under title: Adsorption
and collective paramagnetism.
     Bibliography: p.
     Includes indexes.
     1.   Chemisorption.   2.   Paramagnetism.   I.   Title.
QD547.S44 1975   541'.3453   75-13084
ISBN 0-12-636560-1

PRINTED IN THE UNITED STATES OF AMERICA

Contents

Preface

This work is a complete revision of my book "Adsorption and Collective Paramagnetism" published by Academic Press in 1962. Changes of emphasis have created a need for the revision and the virtual disappearance of the term "collective paramagnetism" in favor of "superparamagnetism" has dictated a change of title. Emphasis now lies more on particle size determination and on the number of adsorbent atoms affected when any molecule is adsorbed on a surface. As before, our concern is chiefly, though not exclusively, with nickel as adsorbent.

This book is addressed principally to readers interested in heterogeneous catalysis and related areas. For that reason a brief introduction to magnetism is included. I also hope that specialists in magnetism and surface physics may find the work of interest.

I am grateful for permission to use material derived from publications of Plenum Publishing Corporation and North-Holland Publishing Company, as indicated in the text. I also wish to acknowledge the permissions and assistance given to me by R. B. Anderson, T. J. Gray, and G.-A. Martin.

P. W. Selwood

Preface

Chemisorption
and Magnetization

I

Introduction

1. Solid–Vapor Interfaces

Adsorption of a molecule on the surface of a ferromagnetic solid produces a change in the magnetization of the solid. If the adsorption process involves appreciable electronic interaction and if the ratio of surface to volume in the adsorbent is large, then the fractional change of magnetization becomes substantial. This lends itself to convenient measurement and to interpretation.

The first report of this effect appears to be that of Morris and Selwood,[1] who found in 1943 that the magnetic susceptibility of nickel supported on (and partially alloyed with) copper suffered a decrease as the system was exposed to carbon monoxide or to hydrogen sulfide. In 1948 Dilke et al.[2] reported that dimethyl sulfide caused a decrease in susceptibility of the paramagnetic metal palladium in the form of powder. The decrease of susceptibility shown by palladium on exposure to hydrogen has been known for over a century, but hydrogen on palladium is not confined to the surface. Dilke et al. related the effect produced by dimethyl sulfide to a true chemisorption involving pairing of d electrons in the metal.

The action of chemisorbed hydrogen in lowering the magnetization of nickel was reported[3] in 1954. It was at first thought that the ferromagnetism of the nickel would make such an effect difficult or impossible to interpret. Actually the unique properties of very small ferromagnetic particles facilitate the interpretation. When the adsorbent possesses unpaired d electrons it is to be expected that formation of a chemical bond may alter the filling of the d band. And if, through bond formation, the adsorbent atom principally involved loses its ability to participate in the cooperative phenomena responsible for ferromagnetism then the result is a loss of magnetization.

It was also reported[3] in 1954 that the specific catalytic activity (for benzene hydrogenation) is dependent on nickel particle size. This conclusion was based on a magnetic determination of particle size involving an assumption that is almost certainly erroneous. But the conclusion concerning specific activity is now known to be valid. The experimental methods used for studying chemisorption and magnetization are, with minor changes, applicable in granulometry. Both areas are presented in detail in later chapters.

The measurement of saturation magnetization in a ferromagnetic substance in the form of small particles is not necessarily easy. Whether the magnetizations at moderate fields and near room temperature bear any simple relation to the saturation magnetization (at infinite field and zero temperature) is something that could scarcely have been predicted prior to development of the Stoner–Néel theory of magnetization in small particles.[4] But Néel's important contribution[5] to the theory was made in 1949, and it is obvious that small particles are required if the adsorbent is to have an appreciable fraction of its atoms in a position to be affected by adsorbate molecules. These conditions are met in a typical nickel–silica hydrogenation catalyst. The method is also applicable to cobalt, to iron, and to a few other adsorbents. There are few restrictions on the kinds of adsorbate for which the method may be used.

2.   Chemisorption

There is no completely satisfactory definition of chemisorption. The least objectionable is probably to say that it is adsorption involving some kind of electronic interaction between adsorbent and adsorbate, but it has been shown that measurable, if slight, interaction occurs for even a molecule such as krypton on nickel. Many systems involve heats of adsorption at least one order larger than the normal heat of condensation of the adsorbate. When this occurs it is clear evidence of chemisorption. But in such systems the heat of adsorption generally falls with increasing surface coverage while, at the same time, other physical and chemical properties may continue to indicate that chemisorption is occurring. It is also to be noted that what appears to be true chemisorption often, indeed generally, occurs at temperatures far above those at which appreciable physical adsorption can be detected.

Physical adsorption is often stated (probably erroneously) to be independent of any specific action on the part of the adsorbent. That is to say, in this view the total volume of vapor taken up by any adsorbent depends, in the case of physical adsorption, solely on the temperature, the pressure, and on the total available surface of the adsorbent. But the situation for chemisorption is quite different. Specific activity of the adsorbent is the rule rather than the exception. In proof of this statement it is sufficient to mention the extensive chemisorption of hydrogen by some metals, of which nickel is one, as contrasted with the negligible chemisorption of molecular hydrogen on copper under identical experimental conditions.

It is doubtful if chemisorption may ever occur beyond the stage of a monolayer of atoms or molecules, as the case may be, on the surface. (This does not mean that an intact molecule may not react with a chemisorbed molecule.) However for physical adsorption, multilayer adsorption is not only possible but at appropriate temperatures will always occur as the pressure of the vapor-phase adsorbate becomes appreciable. Physical adsorption may, of course, occur over a chemisorbed monolayer provided that experimental conditions are appropriate.

It is not infrequently stated that physical adsorption is a rapid process but that chemisorption may be slow. This is a criterion that must be approached with caution. The rate of physical adsorption may, it is true, be limited only by the rate of diffusion to the surface. But if a surface is less accessible by reason of, for instance, being inside a pore of molecular dimensions, then not only will the apparent rate of adsorption be diminished, but Knudsen flow may render the rate almost imperceptible. There is, on the other hand, a mass of evidence tending to show that chemisorption involves, or may involve, an activation energy. This implies that the rate of chemisorption should be measurable. But certainly there are cases in which true chemisorption occurs but for which the rate appears to be instantaneous. It seems to this writer that too often, in the past, the rate of a supposed chemisorption has actually been merely a rate of diffusion or even a relatively straightforward surface reduction as of chromium dioxide by hydrogen at, or near, room temperature.

It may be shown that an adsorbed molecule has suffered some structural change, such as occurs in the hydrogen-deuterium equilibration reaction or in the self-hydrogenation of ethylene on nickel, then it is difficult to escape the conclusion that some electronic interaction of adsorbent and adsorbate, and hence some chemisorption, has taken place. The strictly chemical approaches, including the use of tracers, continue to be among the most powerful methods available to the investigator. To these we must add the application of modern valence theory. To date the conclusions, or rather the verifiable predictions, coming from valence theory have not been spectacular. But there is reason to believe that important developments may lie not too far in the future. Our understanding of chemisorption and of heterogeneous catalysis will not even approach completeness until these developments occur.

There are numerous physical methods that may be used for gaining information concerning the chemisorptive bond. More than a few of these have been discovered or first applied in the last few years. Some of the methods that have yielded new information of major importance are infrared spectroscopy, field emission microscopy, Mössbauer spectroscopy, low-energy electron diffraction, atomic and molecular beam scattering, Auger spectroscopy, electron scanning spectroscopy, and x-ray diffraction. Some of these, together with inelastic-electron tunnelling spectroscopy and photoemission spectroscopy give promise of even more useful information. Reviews of molecular beam and other methods are given by Hobson,[6] Wedler,[7] and Robertson.[8]

It will be noted that some experimental methods give information primarily about the chemisorbed molecule, and some primarily about the possible changes in the surface layer of adsorbent atoms. The magnetic method belongs to the latter group. It is always useful whenever experimental data are available for comparison from two or more different experimental methods for the same or comparable systems. For such comparisons to be valid it is essential that experimenters describe, in full detail, the conditions under which determinations are made.

3.   Definitions in Magnetism

The Gaussian-cgs system of definitions and units for magnetic quantities has been used in this book. This choice was made reluctantly. It was made because of complexities in the Rational (Georgi)-mks system definitions for the quantities field strength H, magnetization, moment, and susceptibility, all of which must be used, not infrequently, in the same equation. Apart from this, SI recommendations for symbols and style have been used throughout. (Compromises of this kind were not unforeseen by IUPAC.) Heavy reliance has been placed on the article by M. L. McGlashan [*Annu. Rev. Phys. Chem.* **24**, 51 (1973)]. Conversion formulas between Gaussian-cgs and Rational-mks units are given in the Appendix.

A magnetic field has both strength and direction. The unit of magnetic field strength is the oersted, Oe. (The gauss, G, is often used instead of oersted. These units are identical.)

Matter placed in a magnetic field becomes magnetized and the magnetic moment m present has the unit $cm^{5/2} \cdot g^{1/2} \cdot s^{-1}$. However we shall find it convenient to use $Oe \cdot cm^3$, although $erg \cdot Oe^{-1}$ is also used. There is no special name for the unit of magnetic moment.

The magnetization M is defined as the magnetic moment per unit volume. Hence the unit of magnetization is the oersted. The oersted has often been used for the quantity $4\pi M$.

The ratio $M/H = \kappa$ is the magnetic susceptibility. It is dimensionless. The magnetic susceptibility divided by the density ρ, that is κ/ρ, has the unit $cm^3 \cdot g^{-1}$. The susceptibility multiplied by the molar volume V_m that is κV_m, has the unit $cm^3 \cdot mol^{-1}$. There are no recommended names or special symbols for κ/ρ or κV_m. The reader is warned that while $M/H = \kappa$ and is dimensionless in both Gaussian and Rational systems, if $\kappa = 1$ in the former it is 4π in the latter.

Many substances have a negative susceptibility and are said to be diamagnetic. Water, with $\kappa/\rho = -0.720 \times 10^{-6}$ $cm^3 \cdot g^{-1}$, is an example of a diamagnetic substance. Specimens of diamagnetic substances are repelled from a region of higher magnetic field to a region of lower field. The susceptibilities are in general, though not always, independent of field strength and of temperature.

Many substances are attracted to a region of higher field from a region of lower field. For these substances the susceptibilities are positive. If the susceptibility is positive, independent of field strength, and varies inversely, or nearly so, with thermodynamic temperature the substance is said to be paramagnetic. A typical paramagnetic substance is $CuSO_4 \cdot 5H_2O$ for which, at 293 K, $\kappa/\rho = 5.85 \times 10^{-6}$ $cm^3 \cdot g^{-1}$.

For most paramagnetic substances below a characteristic critical temperature there is an abrupt, and often large, change of susceptibility. The susceptibility, though still positive, may become dependent on field strength and virtually independent of temperature. If, on passing down through the critical point the susceptibility greatly and abruptly increases, the substance is said to be ferromagnetic. An example of a ferromagnetic substance is metallic nickel for which the critical temperature T_C (called the Curie point) is near 631 K. But if, on passing down through the critical point the susceptibility decreases, the substance is said to be antiferromagnetic. An example of an antiferromagnetic substance is α-Cr_2O_3 for which the critical temperature T_N (called the Néel point) is at 307 K.

Other more complicated kinds of magnetic behavior are known. In this book we shall be concerned with ferromagnetism, with paramagnetism, and with the borderline between them. These kinds of matter contain permanent magnetic moments arising from unpaired electrons. Diamagnetic forms of matter have no permanent electronic moments but they have induced moments when placed in a magnetic field. This is true also of ferromagnetic and paramagnetic matter but the correction for diamagnetism is negligible for most of our purposes.

4. Paramagnetism

Let m_p be the magnetic moment of uniform particles of any kind. In a sample containing n_p moles of particles there will be $n_p L$ particles, where L is Avogadro's constant. If, in an applied field H orientation of these moments is complete in the direction of the field the magnetization will reach saturation. This is called the saturation magnetization M_s. But M is less than M_s in paramagnetic matter except at very high field and very low temperature. The ratio M/M_s is given by the Langevin equation:

$$\frac{M}{M_s} = \frac{\coth m_p H}{kT} - \frac{kT}{m_p H} \tag{1.1}$$

where T is the thermodynamic temperature and k is the Boltzmann constant.

If $M \ll M_s$ the Langevin equation may be written

$$M = n_p L \, m^2{}_p \, H/3kt \tag{1.2}$$

and this may be rewritten

$$\kappa = n_p L m_p{}^2/3kT \tag{1.3}$$

Equation (1.3) is in agreement with the Curie law $\kappa = C/T$, where C is called the Curie constant. More often it will be found that $\kappa = C/(T + \Delta)$, which is called the Curie–Weiss law. The Weiss constant (often written $-\Delta$) has only limited theoretical significance.

Equation (1.3) may be used to obtain m_p from susceptibility data. (These should be measured over a range of temperature and, if it is found that the data are better represented by the Curie–Weiss law, $T + \Delta$ should be substituted for T.)

Then we may write

$$m_p = \left(\frac{3kT\kappa}{n_p L}\right)^{1/2} \tag{1.4}$$

or, in a sample of unit volume equal to V_m,

$$m_p = \left(\frac{3kT\kappa V_m}{L}\right)^{1/2} \tag{1.5}$$

We now calculate m_p for the case in which the particles are individual nickel ions in aqueous solution. From measurements[9] on $NiCl_2$ dissolved in water at 293 K the value of $\kappa V_m = 4.43 \times 10^{-3}$ cm³·mol⁻¹ for the Ni^{2+} after appropriate diamagnetic corrections. The Weiss constant is zero.

$$m(Ni^{2+}) = \frac{3 \times 1.38 \times 10^{-16} \text{ cm}^2 \cdot \text{g} \cdot \text{s}^{-2} \cdot \text{K}^{-1}}{6.022}$$

$$\times \frac{2.93 \times 10^2 \text{ K} \times 4.43 \times 10^{-3} \text{ cm}^3 \cdot \text{mol}^{-1}}{10^{23} \text{ mol}^{-1}}$$

$$= 2.95 \times 10^{-20} \text{ cm}^{5/2} \cdot \text{g}^{1/2} \cdot \text{s}^{-1}$$

$$= 2.95 \times 10^{-20} \text{ Oe} \cdot \text{cm}^3$$

It is convenient to express moments such as the above by the ratio m_p/m_B, where $m_B = eh/4\pi m_e$. The quantity m_B is the Bohr magneton, with e and m_e the charge and rest mass, respectively, of an electron and h is Planck's constant. As

$$m_B = 0.927 \times 10^{-20} \text{ erg} \cdot \text{Oe}^{-1} = 0.927 \times 10^{-20} \text{Oe} \cdot \text{cm}^3$$

we have $m(Ni^{2+})/m_B = 3.2$. This is called the Bohr magneton number for Ni^{2+}. It is dimensionless. We shall use the symbol β for the Bohr magneton number. Then β for any atom is easily found from the experimental susceptibility as follows:

$$\beta = 2.84(\kappa V_m T)^{1/2} \qquad (1.6)$$

Magnetic moments depend not only on the number of unpaired electrons in the atom or ion but also on the way in which the spin and orbital angular momenta may be combined. If only the spin component need be considered, the Bohr magneton number is given by

$$\beta = 2[S(S+1)]^{1/2} \qquad (1.7)$$

where S is the spin quantum number equal to half the number of unpaired spins in the atom. Equation (1.7) may be applied to a large number of substances with success. It is known as the "spin-only" formula. Table I shows experimental and calculated Bohr magneton numbers for several ions. It must be understood that these numbers are not observed for all compounds in which the ions may be present. For most of the paramagnetic ions shown the orbital contribution is said to be "quenched." But for many substances, such as Co^{2+}, the observed moment is better represented by

$$\beta = g[J(J+1)]^{1/2} \qquad (1.8)$$

where g is the Landé splitting factor and J the total quantum number. Equation (1.8) is especially useful for predicting the moments of the

TABLE I

MAGNETIC MOMENTS OF IONS OF THE FIRST TRANSITION SERIES

Ion	3d electrons	Unpaired electrons	$2[S(S+1)]^{1/2}$	β (obs)
Sc^{3+}, Ti^{4+}, V^{5+}	0	0	0.00	0.0
Ti^{3+}, V^{4+}	1	1	1.73	1.8
V^{3+}	2	2	2.83	2.8–2.9
V^{2+}, Cr^{3+}, Mn^{4+}	3	3	3.87	3.7–4.0
Cr^{2+}, Mn^{3+}	4	4	4.90	4.8–5.1
Mn^{2+}, Fe^{3+}	5	5	5.92	5.2–6.0
Fe^{2+}	6	4	4.90	5.0–5.5
Co^{2+}	7	3	3.87	4.4–5.2
Ni^{2+}	8	2	2.83	2.9–3.4
Cu^{2+}	9	1	1.73	1.8–2.2
Cu^{+}, Zn^{2+}	10	0	0.00	0.00

lanthanides and the actinides where, for the most part, the f electrons are shielded from external electric fields and other effects. But the presence of adjacent atoms or ions, and their geometric arrangement, often cause substantial changes in the observed magnetic moments, as does the Heisenberg exchange interaction that becomes important below the critical temperature in all paramagnetic substances. This area, especially with respect to metal oxides, is described in detail by Schieber.[10] The susceptibility times specific

TABLE II

SUSCEPTIBILITY DATA, κ/ρ cm$^3\cdot$g^{-1}, AT ROOM TEMPERATURE

Substance	$(\kappa/\rho) \times 10^6$	Substance	$(\kappa/\rho) \times 10^6$
Alumina, Al_2O_3	−0.3	Magnesium, Mg	0.25?
Aluminum, Al	+0.6	Mercury, Hg	−0.17
Argon, Ar	−0.5	Nitrogen, N_2	−0.43
Calcium, Ca	+0.7	Oxide ion, O^{2-}	−0.75
Copper, Cu	−0.83	Oxygen, O_2	+107.8
Graphite, C	−7.8[a]	Platinum, Pt	+1.0
Gold, Au	−0.15	Silica, SiO_2	−0.5
Hydrogen, H_2	−2.0	Silver, Ag	−0.2
Hydroxide ion, OH^-	−0.70	Zinc, Zn	−0.17

[a] Crystalline graphite shows a remarkable example of magnetic anisotropy. The susceptibility of the powder, which is actually an average of the three principal susceptibilities as measured along the magnetic axes, is strongly dependent on particle size. The value given is for coarsely powdered crystals.

volume, κ/ρ, is given for a few substances of interest for our purpose in Table II.

5. Ferromagnetism

As previously mentioned some substances including iron, cobalt, nickel, various oxides, and other compounds become magnetized to a very large degree if placed in a magnetic field of quite moderate strength. But as the applied field is raised the magnetization reaches a limit as shown in Fig. 1. Matter acting in this manner is said to be ferromagnetic. Such forms of matter also show unique behavior as the temperature is raised. As the temperature approaches a value characteristic of each substance the magnetization falls abruptly. This is called the Curie temperature. Some distance above T_C the substance may act as a typical paramagnetic substance with susceptibility described by the Curie–Weiss law. Sometimes the Weiss constant is about the same numerically as the thermodynamic Curie temperature.

The reason that ferromagnetic matter acts in the above manner is that below T_C such substances are actually magnetized even in the absence of an external field. It was first suggested by Pierre Weiss that in ferromagnetic matter there is a large internal field (of the order of 10^6 Oe), that this field is proportional to the magnetization M, and that it causes the magnetization to approach M_s corresponding to complete parallel orientation of the atomic moments. In paramagnetic matter the orientation of magnetic dipoles is opposed by thermal agitation. But in ferromagnetics electron

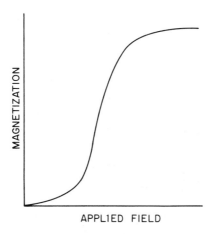

FIG. 1. Magnetization as a function of field for typical ferromagnetic matter.

spin moments are locked together in the same sense so that they act co-operatively. The group acts as one very large magnetic moment. In such groups the electron spins are held in parallel alignment by quantum-mechanical (Heisenberg) exchange forces. This parallel alignment persists against thermal agitation until it finally breaks down above the Curie temperature.

It may be wondered why it is that a piece of iron with $T_C = 1043$ K may readily be "demagnetized" by appropriate treatment. The reason for this is that the parallel orientation of spin moments in ferromagnetic matter occurs in small volumes called Weiss domains. Within each domain the spin moments are parallel at all temperatures below T_C. But orientation of the resultant moment may be quite different in different domains, even though these domains may be adjacent to each other. It may, therefore, occur that the overall resultant magnetization of any particular specimen may be small if the direction of magnetization in different domains is random. These domains may contain only a few atoms, or they may be large enough to observe under low magnification. Techniques are available for making them visible and for determining the direction of magnetization.

In the presence of an applied field the direction of magnetization tends to turn in the direction of the field. This turning may be coherent within each domain or it may involve growth of properly oriented domains at the expense of those with resultant moments pointed in different directions. This wholesale alignment is the process by which the sample is said to become "magnetized." The limit is reached when orientation is effectively complete.

The magnetization so obtained, even though the applied field need not be very large, is often called the "saturation" magnetization. It does, how-ever, vary with temperature and, to some degree, with field strength. We shall use the term "spontaneous magnetization" with the symbol M_{sp} and unit Oe.*

The spontaneous magnetization of iron at room temperature is 1707 Oe, at 0 K it is 1752 Oe. The variation with temperature which is of much the same form for many ferromagnetics is shown in Fig. 2, in which relative spontaneous magnetization· is plotted against T/T_C. The true saturation magnetization is the magnetization at infinite field and absolute zero. This will be designated M_0. Complete alignment of atomic magnetic dipoles at any temperature other than absolute zero may be achieved, at least in principle, at infinite field. This will produce the saturation magnetization designated M_s.

---

* Some authors restrict the term spontaneous magnetization to the magnetic moment per unit volume within a domain at temperature T and field zero. We shall find that the definition given above will be satisfactory for our purposes.

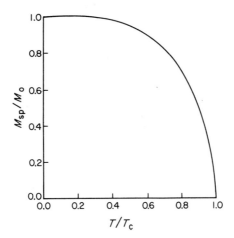

FIG. 2. Relative spontaneous magnetization versus reduced Curie temperature T/T_C for typical ferromagnetic matter.

We shall now calculate the Bohr magneton number for iron, $\beta(\text{Fe})$, from saturation data at 0 K. It will be recalled that the magnetization $\boldsymbol{M}$ is the magnetic moment $\boldsymbol{m}$ per unit volume. Then in a sample of unit volume containing n_{Fe} moles of iron there will be $n_{\text{Fe}}L$ atoms and

$$\boldsymbol{M}_0 = n_{\text{Fe}}\boldsymbol{m}(\text{Fe})L \tag{1.9}$$

where $\boldsymbol{m}$ is the "saturation" moment of each identical particle, namely, each atom. Hence

$$\boldsymbol{m}(\text{Fe}) = \boldsymbol{M}_0 V_\text{m}/L \tag{1.10}$$

where V_m is, as before, the molar volume. For iron $\boldsymbol{M}_0 = 1752$ Oe, $\rho = 7.895$ g·cm^{-3}, and the molar mass is 55.85 g·mol^{-1}, we have therefore

$$\boldsymbol{m}(\text{Fe}) = \frac{1.753 \times 10^3 \text{ Oe} \times 5.585 \times 10 \text{ g·mol}^{-1}}{7.895 \text{ g·cm}^{-3} \times 6.022 \times 10^{23} \text{ mol}^{-1}}$$

$$= 2.06 \times 10^{-20} \text{ Oe·cm}^3$$

The Bohr magneton number for iron, $\beta(\text{Fe})$, is then

$$\frac{2.06 \times 10^{-20} \text{ Oe·cm}^3}{0.927 \times 10^{-20} \text{ Oe·cm}^3} = 2.22$$

This is dimensionless.

Some Bohr magneton numbers (β) and Curie temperature T_C (K) are given in Table III.

TABLE III

BOHR MAGNETON NUMBERS AND CURIE TEMPERATURES
FOR SOME FERROMAGNETICS

Substance	β	T_C (K)
Fe	2.22	1043
Co	1.7	1403
Ni	0.6	631
Gd	7.12	289
EuO	6.9	77

The saturation moment so obtained should not be confused with the paramagnetic moment discussed in the previous section. The paramagnetic magneton number is given by $g[J(J + 1)]^{1/2}$ or, if we may consider electron spins only, by $2[S(S + 1)]^{1/2}$. But the saturation moment is simply $2S$. The reason for this difference is that the paramagnetic moment is the actual moment, but the saturation moment is the maximum moment resolved parallel to the applied field.

There is at the present time no completely satisfactory theory concerning the arrangement of electrons in d metals. This paragraph is inserted in an effort to make intelligible some of our later remarks. A current theory of electron distribution is known as the band theory. It is thought that the energy levels need not be localized around atoms or ions in the familiar sense, but that they are rather spread out over the whole crystal mass. The isolated atoms of iron, cobalt, and nickel have, respectively, 6, 7, and 8 electrons in the d shell. From this the magneton numbers might be expected to be 4, 3, and 2, respectively. Actually the "saturation" numbers are 2.22, 1.7, and 0.6. (The "paramagnetic" numbers from susceptibility measurements well above T_C are about 3.5, 3.1, and 1.7.) According to band theory it is assumed that the 3d states and the 4s states overlap in such a way that, on the average, fractional filling of the several states is possible. To explain, for instance, the value $\beta(\text{Ni}) = 0.6$, it is assumed that of the 10 available 3d states electrons actually occupy 9.4. This leaves 0.6 unopposed spins per atom. The remaining 0.6 electron is, on the average, to be found in the 4s band, but s electrons are thought to make no contribution to the magnetic moment.

The final topic to which we shall refer in this section is magnetic anisotropy. In general a sample of crystalline matter tends to orientate in a uniform magnetic field. The only exceptions are crystals such as the cubic of high symmetry or polycrystalline matter in which there is no preferred direction. The reason for this behavior in diamagnetic and paramagnetic solids is that the susceptibility of the molecule may be different in different directions. Crystalline benzene is an example in which the two principal values of κ/ρ parallel to the ring are about -0.5×10^{-6} while the one perpendicular to the ring is about -2.5×10^{-6}. In ferromagnetic solids the phenomenon of magnetic anisotropy may be of prime importance. Not only may the saturation moments be somewhat different along different axes but, of considerably more practical importance, it may be much easier to magnetize the crystal in certain directions. Ferromagnetic anisotropy may arise simply because of the shape of the crystal. A needle-shaped crystal is much easier to magnetize, generally, in the direction of the long axis than it is at right angles to this axis. (This effect is also present in diamagnetics and paramagnetics but it is not nearly so pronounced.) But ferromagnetic anisotropy may also arise from the arrangement of atoms in the crystal. Cobalt is, for instance, easy to magnetize along the hexagonal axis but hard to magnetize at right angles to this axis. This phenomenon is known as magnetocrystalline anisotropy; it apparently arises from the effect of electrostatic fields produced by the atoms themselves on the spin-orbital coupling and partial quenching of the orbital contribution.

There are several other sources of anisotropy. One of the most important is strain anisotropy resulting, as the name implies, from dislocations produced by mechanical strains in the sample. These strains may arise from mechanical working of the sample, but they often appear to an exaggerated degree in small particles or in thin-metal films formed by condensation from the vapor. We shall have reference to magnetic anisotropy later. One consequence is that anisotropy has an influence during demagnetization just as it does during magnetization. If, therefore, an anisotropic substance is first magnetized and then the external field is reduced to zero it will be found that the sample still shows a more or less strong magnetization. This means that the spontaneous magnetization in the domains continues to have a preferred direction which may be overcome only by applying an appropriate field in the reverse direction, or by heating the sample above its Curie temperature. These effects give rise, as is well known, to the phenomena of magnetic hysteresis, remanent magnetization, and coercive force.

6. Antiferromagnetism and Ferrimagnetism

In many substances for which constituent atoms have unpaired electrons the Heisenberg exchange interaction is negative rather than positive, as it is in ferromagnetic substances. Negative interaction leads to diminishing susceptibility below the phase transition temperature called the Néel point, T_N. Sometimes the transition from paramagnetic to antiferromagnetic is accompanied by an abrupt change of susceptibility as in MnO, $T_N = 122$ K, but for α-Cr$_2$O$_3$, $T_N = 310$ K, the change is gradual although the rate of the catalyzed parahydrogen conversion over this oxide shows an increase of almost one order within a degree below T_N.

Most oxides of transition elements, and many other compounds, exhibit antiferromagnetism at the appropriate temperature. The phenomenon is more common than is ferromagnetism. For antiferromagnetism to occur it is not necessary that the atoms with unpaired spins should be adjacent to each other as in metals (nor is it always necessary for ferromagnetic interaction to occur). Interaction through an intervening ion, such as O^{2-}, is called superexchange. It is of some interest to workers in the field of heterogeneous catalysis that supported oxides such as Cr$_2$O$_3$ on high-area Al$_2$O$_3$ and also the familiar high specific surface gels such as chromia aerogel, tend to be normal paramagnetics rather than antiferromagnetics. The reason for this is doubtless the degree of attenuation which places each ion in an environment with much less than its normal coordination of neighboring atoms. Antiferromagnetics do not have the domain structure characteristic of ferromagnetics.

There is a fairly large group of substances of which many occur in the spinel structure. Those called ferrites may be represented by the general formula MO·Fe$_2$O$_3$. The metal M may be, for example, Zn^{2+}, Mn^{2+}, or Fe^{2+}. Various combinations are possible, part of the (3+) iron may be replaced by other elements, part of the oxygen by sulfur. Combinations including rare earths, known as rare-earth garnets, are also well known. In these substances some of the electron spins may be parallel (as in ferromagnetics) and some may be antiparallel (as in antiferromagnetics). This circumstance, first described by Néel, makes it possible to explain, and predict, the complicated magnetic behavior often shown. For instance, in Fe$_3$O$_4$ which is an inverse spinel better written Fe^{2+}Fe$_2^{3+}$O$_4$, for each Fe^{3+} in a tetrahedral hole there is one Fe^{3+} and one Fe^{2+} in octahedral holes. The saturation Bohr magneton numbers are, 5 for Fe^{3+} and 4 for Fe^{2+}. In each Fe$_3$O$_4$ group there are five electron spins in tetrahedral coordination directed antiparallel to five spins in octahedral coordination. This leaves four spins in octahedral coordination able to contribute to the magnetization. The observed β(Fe$_3$O$_4$) is about 4.2. But if the Fe^{2+} ions are progressively replaced by diamagnetic Zn^{2+} ions the Bohr magneton number falls to zero. Néel suggested the name ferrimagnetism for this phenomenon.

There are many other kinds of magnetic phenomena involving cooperative effects between electron spins. These are described by Schieber.[10] In this book we shall have a little more to say about ferrimagnetism, but not about antiferromagnetism.

References

1. H. Morris and P. W. Selwood, *J. Amer. Chem. Soc.* **65,** 2245 (1943).
2. M. H. Dilke, D. D. Eley, and E. B. Maxted, *Nature (London)* **161,** 804 (1948).
3. P. W. Selwood, S. Adler, and T. R. Phillips, *J. Amer. Chem. Soc.* **76,** 2281 (1954); **77,** 1462 (1955).
4. E. C. Stoner, *Phil. Trans. Roy. Soc. London Ser. A* **235,** 165 (1936).
5. L. Néel, *Ann. Géophys.* **5,** 99 (1949).
6. J. P. Hobson, *Jap. J. Appl. Phys., Proc. Int. Vac. Congr. Kyoto, 6th, 1974.*
7. G. Wedler, "Adsorption." Verlag Chemie, Weinheim, 1970.
8. A. J. B. Robertson, "Catalysis of Gas Reactions by Metals." Logos Press, London, 1970.
9. H. R. Nettleton and S. Sugden, *Proc. Roy. Soc. London* **A173,** 313 (1939).
10. M. M. Schieber, "Experimental Magnetochemistry." North-Holland Publ. Co., Amsterdam, Wiley, New York, 1967.

II

Superparamagnetism

1. Very Small Particles

The presence of a layer of adsorbed molecules on the surface of a metal could hardly be expected to cause a measurable change in the magnetization unless 1% or more of the metal atoms were on the surface. Reasonable precision might be expected if 10% or more were so situated. This requirement means that the metal particles must be rather less than 10 nm in diameter and must, therefore, contain no more than a few thousand atoms. It is a fortunate circumstance that the nickel particles in a typical nickel–silica hydrogenation catalyst average about 5 nm in diameter, or even less. Commercial nickel catalysts are found, not infrequently, to be suitable for magnetic investigation of chemisorption processes.

A very small particle of ferromagnetic matter may be essentially a single magnetic domain. Such particles exhibit properties that are unique and that lie on the borderline between ferromagnetism and paramagnetism. Michel[1] showed in 1937, and more specifically[2] in 1950, that the slow reduction of nickel–silica preparations may yield nickel in a form that has certain aspects of ferromagnetism but the magnetization of which, at constant field, increases markedly with decreasing temperature far below the normal Curie point. Michel's interpretation of this anomaly was that very small particles of ferromagnetic substances should have Curie points lower than that of the same substance in massive* form, and that typical preparations containing a wide range of particle diameters would possess a wide range of Curie points, of which none would be well defined. This interpretation is almost certainly of minor significance insofar as it con-

---

* The word "massive" is used here to mean matter in the form of relatively large, well-crystallized pieces. Some authors use the term "bulk" to mean the same thing.

cerns the the magnetization–temperature curves of reduced nickel–silica preparations. But Michel's ideas guided some of the early work on the subject and even now it is not certain what effect particle size has on the Curie temperature. We shall return to this point later.

The view that a particle of a ferromagnetic substance, below a certain critical size, would consist of a single domain was suggested by Frenkel and Dorfman.[3] The term "single domain" may have several meanings. We shall use it to mean a particle which is in a uniform state of magnetization at any applied field. Such a particle may have a diameter of 30 nm or less depending on the particular substance. These particles may exhibit a kind of magnetic Brownian movement in such a way that orientation of the magnetic moment of the particle considered as a whole is affected by thermal agitation. The particle, when placed in an external field, tends to behave like a paramagnetic atom, but one that has a very large magnetic moment. That such behavior actually occurs was shown by Elmore[4] who studied colloidal suspensions of magnetite and of γ-ferric oxide. But there is also a mechanism of thermal relaxation not involving physical rotation of the particle.[5]

Several names have been suggested for the phenomenon outlined above. "Superparamagnetism" is now almost universally used.*

The first treatment of superparamagnetism appears to be that of Stoner[6] which was based in part on earlier considerations of Gans and Debye.[7] The magnetization M of an assembly of single domain particles, each consisting of not more than a few thousand atoms, may be described by the Langevin equation (1.1). Recalling, again, that the magnetization is the moment per unit volume we may write

$$m_\mathrm{p} = M_\mathrm{sp} v \tag{2.1}$$

where v is the volume of a particle. Hence, for an assembly of uniform particles,

$$M = M_\mathrm{sp} V \left(\coth \frac{M_\mathrm{sp} v H}{kT} - \frac{kT}{M_\mathrm{sp} v H} \right) \tag{2.2}$$

V being the volume of the sample.

It might be thought that, as is the case for paramagnetics, it would be necessary to use the Brillouin, rather than the Langevin, function to describe the magnetization at very low temperatures, but a particle containing several hundred atoms may be thought of as having a spin quantum

---

* The only objection to "superparamagnetism" is that it was previously used by Kobozev for another effect [N. I. Kobozev et al., J. Phys. Chem. USSR **33**, 641 (1959) (English transl.)].

number S in the hundreds. Such a particle is adequately described by classical theory. It will be noted also that the difference between paramagnetic and saturation moments vanishes. This difference, which is the difference between $2[S(S+1)]^{1/2}$ and $2S$, becomes negligible as S becomes quite large.

For all real adsorbent samples to be discussed here the particle volume is far from uniform. Let us assume a distribution of particle volumes, $f(v)$, where

$$\int_0^\infty f(dv) = 1 \tag{2.3}$$

Then

$$\boldsymbol{M} = V \int_0^\infty \boldsymbol{M}_{\text{sp}} \left[\coth\left(\frac{\boldsymbol{M}_{\text{sp}} v \boldsymbol{H}}{kT}\right) - \frac{kT}{\boldsymbol{M}_{\text{sp}} v \boldsymbol{H}} \right] f(v) \, dv \tag{2.4}$$

If $\boldsymbol{M}_{\text{sp}}$ is independent of v

$$\boldsymbol{M} = V \boldsymbol{M}_{\text{sp}} \int_0^\infty \left[\coth\left(\frac{\boldsymbol{M}_{\text{sp}} v \boldsymbol{H}}{kT}\right) - \frac{kT}{\boldsymbol{M}_{\text{sp}} v \boldsymbol{H}} \right] f(v) \, dv \tag{2.5}$$

This equation is in a form that permits correction for variations of $\boldsymbol{M}_{\text{sp}}$ with temperature. Some authors use the magnetization multiplied by specific volume $\sigma = \boldsymbol{M}/\rho$ instead of $\boldsymbol{M}$ in the above equations.

The treatment given, which is primarily from the papers by Bean[8,9] and his associates, is based on the assumption that no remanence is ever present. This assumption is often justifiable at room temperature and above but it is never justifiable for any catalyst thus far studied at very low temperature. The reasons for this have an important bearing on our overall problem. They are discussed in the following section.

2. Anisotropy and Relaxation

Some geological deposits of ferrimagnetic iron oxide are found to be magnetized in a direction not readily related to the present direction of the Earth's magnetic field. As part of an attempt to explain this anomaly Néel[5] developed a theory of magnetization in small particles. Further progress and applications were made by Bean and his associates[8] in connection with a study of dilute solid solutions of cobalt in copper metal.*

* It is remarkable that a development in geophysics plus one in the precipitation hardening of metals should have applications in heterogeneous catalysis.

Real particles are never truly isotropic. Let there be a particle of moment m_p directed at an angle θ to an applied field H. If the anisotropy of the particle is uniaxial, the anisotropy contribution to the total energy may be

$$E_a = K'v \sin^2 \theta \tag{2.6}$$

where E_a is the anisotropy energy, K' the anisotropy constant, v the volume of a particle, and where the Boltzmann distribution of the angles θ to the field will be different than if the particle were isotropic.

In general, the anisotropy energy is proportional to the volume of the particle. As shown previously, anisotropy may arise from various sources. Large particles, or elongated particles, may deviate substantially from behavior analogous to true paramagnetism, or in other words they may no longer be described in terms of superparamagnetism. Such an assembly of particles may be magnetized, but if the magnetizing field is removed, the magnetization will diminish in a time that is finite but one that may be very long. Néel showed that the remanence M_r at time t is given by

$$M_r = M \exp(-t/\tau) \tag{2.7}$$

where M is here the magnetization at $t = 0$ and τ, the relaxation time, is given by

$$1/\tau = f_0 \exp(-K'v/kT) \tag{2.8}$$

f_0 being a frequency factor of the order of 10^9 s^{-1}. Random orientation after removal of the magnetizing field, or reorientation after change in direction of the field, thus involves an activation energy. One solution to Néel's original problem concerning magnetic rocks is that the deposits were laid down, in ancient times, parallel to the Earth's field, but that the positions of the Earth's magnetic poles have changed. The relaxation time for these particles is, therefore, in the megayear range.

Some idea of particle volumes and relaxation time may be found from Eqs. (2.7) and (2.8) with the aid of anisotropy constants given by Boz-

TABLE IV

MAGNETIC ANISOTROPY CONSTANTS (erg·cm^{-3} = 10^{-1} J·m^{-3})
FOR IRON, COBALT, AND NICKEL

Temperature (K)	Fe (bcc) $K_1' \times 10^{-3}$	Co (hcp) $(K_1' + K_2') \times 10^{-6}$	Ni (fcc) $K_1' \times 10^{-3}$
4.2	575	9	-750
77	560	9	-650
300	480	6	-35

TABLE V

CRITICAL RADII[a] FOR DECAY OF M_s TO M_r

Metal	Critical radii (nm)		
	300 K	77 K	4.2 K
Fe (bcc)	12.7	7.7	2.9
Co (hcp)	4.8	1.9	0.7
Ni (fcc)	43.8	10.5	3.8

[a] Based on relaxation time, $\tau = 10$ s, and decay to 1% in about 0.4 ks.

orth[10] as in Table IV. For our purposes, we shall simply point out that $K'v$ is a measure of the energy barrier over which the direction of magnetization in the particles has to be reversed by thermal activation.

To simplify the calculation the method followed will be that of Bean and Livingston.[8] This is to consider a relaxation time of $\tau = 10^2$ s to be a criterion of superparamagnetism. (Perhaps a more realistic criterion would be to consider τ to be short compared with the time necessary for any particular experiment.) We shall also consider that the energy barriers along certain crystallographic axes have the following values: $\frac{1}{4}K'v$ for $K' > 0$, ([100] easy direction); and $\frac{1}{12}K'v$ for $K' < 0$, ([111] easy direction). Cobalt is a rather special case for which the barrier is taken[11] as $(K_1' + K_2')v$.

If $\tau = 10^2$ s, then $v \simeq 25 \, kT/K_1'$ from which the radii given in Table V may be found. These are the radii of spherical particles, calculated with certain simplifying assumptions, for which M_r will decay in about 6 or 7 min to 1% of its value at $t = 0$. If the particles are not spherical the decay time will be longer.

While the radii given in Table V are approximate only, they give some idea of the magnitudes, and they show that it requires a smaller particle of cobalt than of nickel to exhibit superparamagnetism under the same conditions. It will also be noted that the rate at which M_r decays is very sensitive to particle radius. These topics will be developed in more detail in Chapter IV.

3. Experimental Evidence for Superparamagnetism

A consequence of Eq. (2.2) is that magnetizations, or M/M_s, obtained at different temperatures may be superimposed if plotted with respect to H/T. Plots of this kind are often used as evidence of superparamagnetism. A correction must be made for the change of M_{sp} with T.

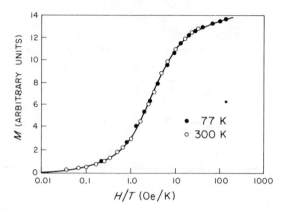

FIG. 3. Superparamagnetism shown by precipitated cobalt particles in a copper matrix (after Becker, Ref. 12).

An example of superparamagnetism is given by Becker[12] who investigated small particles of cobalt in copper. The procedure was to quench a 2% cobalt solid solution from 1323 K. The alloy was then heated briefly at 923 K and quenched again. This procedure caused precipitation of cobalt particles averaging only 2.4 nm in diameter. Figure 3 shows the H/T superposition curve as obtained at 77 and 300 K.

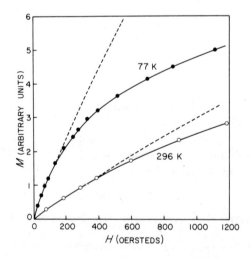

FIG. 4. Field strength dependence of magnetization shown by a reduced nickel–silica catalyst prepared by coprecipitation.

Similar studies have been made by many investigators. We shall describe in some detail the results obtained by Dietz and Selwood[13] on a nickel preparation of the kind familiar in heterogeneous catalysis. This was prepared by coprecipitating nickel hydroxide and silica by mixing boiling solutions of nickel nitrate and basic sodium silicate as described by van Eijk van Voorthuysen and Franzen[14] (their preparation CLA-5421). This was dried, compressed into pellets, and finally reduced *in situ* in flowing hydrogen at 623 K for 12 hr. The hydrogen was removed by evacuation before the temperature was lowered, and it was replaced by purified helium before the magnetization was measured.

Figure 4 shows M (in arbitrary units) plotted against H at 77 and 296 K for the above sample. It will be noted that at relatively low values of H/T the susceptibility, $M/H = \kappa$, is nearly constant as predicted by Eq. (1.3) for a paramagnetic sample.

Figure 5 shows the data of Fig. 4 replotted as M versus H/T and after appropriate corrections for the demagnetizing field (see Chapter III) and the change of M_{sp} with temperature (p. 36). The superposition of points shown in Fig. 5 is satisfactory although less so at higher values of H/T. We may say that the sample exhibits superparamagnetism at room temperature and at fields of up to about 1 kOe. An example of strong deviations from the superposition test is shown in Fig. 6. This is for a nickel–silica preparation that has been heated, in hydrogen, above 673 K and that consequently contains larger nickel particles formed by sintering.

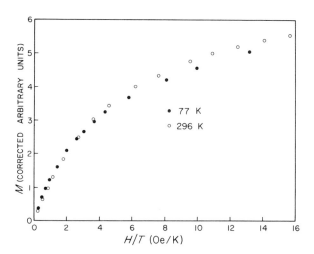

Fig. 5. The data of Fig. 4 replotted to show M as a function of H/T after appropriate corrections for demagnetizing fields and the change of M_{sp} with T.

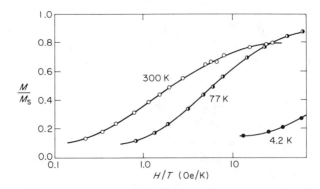

Fɪɢ. 6. Effects due, presumably, to anisotropy in preventing M versus H/T super-position for a nickel–kieselguhr sample that had been sintered and that consequently contained particles larger than those used in obtaining the data shown in Fig. 5.

4. Particle Size and Saturation Moment

In this section there will be described a possible major complication in the interpretation of magnetization measurements on very small parti-cles. This involves the assumption that M_s is constant over the range of particle diameters encountered. There are some theoretical reasons for suspecting that a clean, smooth surface of nickel, and perhaps of some other ferromagnetic substances, might be paramagnetic as is the case for pal-ladium, or possibly even nonmagnetic.[15] Surface layers in which the latter possibility occurs are sometimes referred to as "dead layers," although this term is singularly inappropriate if we are thinking in terms of catalytic activity rather than of magnetism. If dead layers actually exist then our attempts to use magnetic measurements in particle size determination become extremely complicated, and the magnetic study of chemisorption becomes even more obscure.

The most important experimental evidence favoring dead layers is that of Liebermann and Clinton[16] who have measured the magnetization of very thin nickel films and who find that the first few atomic layers are non-magnetic, after which the magnetization rises linearly with film thickness. It is also found that for a given film the number of (atomic) dead layers increases with increasing temperature, apparently reaching a minimum of two at low temperatures and rising to 10 or 20 at 513 K. This temper-ature dependence is reversible. Similar observations by Liebermann *et al.* have been made on iron and on cobalt films, but in these metals only two layers remain dead.

It is true that the familiar catalytic preparations of nickel rarely involve surfaces that are smooth. First, we shall refer to other magnetization studies on nickel films in which at least partial smoothness is almost certainly present and in which surface cleanliness is not suspect. Neugebauer,[17] working with vacuum-deposited films, found saturation magnetizations as measured at room temperature to be independent of film thickness down to 2.0 nm. He states that a ferromagnetic dead layer, if present, can be no more than one-half atom layer thick. Walker et al.[18] studied vacuum-deposited iron films (overcoated with silver) in the thickness range 1.0–10.0 nm, and over a range of temperature. Investigation by Mössbauer spectroscopy showed no structure that could be attributed to dead layers.

Turning to small particles we find that Bean and Livingston[19] report no change greater than 2% in the saturation magnetization of cobalt down to a diameter of 2.1 nm. The cobalt was precipitated from solid solution in copper. Yet if a 2.1 nm particle had two atomic dead layers it would lose about one-half its magnetization. Comparable results were obtained for iron particles in mercury. A study by Takajo et al.[20] of fine particles of nickel obtained by vaporization at low pressure in helium or argon yielded diameters in the 3.5–20.0 nm range. These particles showed no evidence of diminished magnetization. On the other hand, Shinjo et al.,[21] using a film preparation method similar to that of Liebermann, report some shrinkage of the surface magnetization in cobalt.

There have been several studies of saturation magnetization in small particles of nickel prepared by coprecipitation or impregnation, followed by reduction, on silica. These have, without exception, shown no dependence of M_s on particle size. Such studies require independent and sometimes indirect estimates of total nickel and of metallic nickel present. Detailed consideration of these results will be deferred until a later chapter.

To reconcile the results of Liebermann and Clinton with the conflicting earlier work, it is necessary to consider in more detail the experimental method used by these authors in the preparation of nickel films. The films were deposited by electrolysis from aqueous solution. It seems most probable that chemisorbed water is responsible for lowering the magnetization. It will be shown in later chapters that all chemisorbed molecules cause a loss of magnetization proportional to the surface coverage. Even if the nickel film is removed and dried it would still hold a monolayer of water that would progressively change to hydroxide ion and then to oxide ion as the temperature is raised. No method short of chemical reduction at relatively high temperature would remove a layer of oxide so formed.

In view of these considerations we conclude that no influence of particle size on saturation magnetization (within the range under consideration) has yet been demonstrated.

5. Particle Size and Internal Field

There is another possible complicating effect of particle size on magnetizations. This is related to some uncertainty concerning the Curie temperature T_C and whether or not it may be different for small particles of ferromagnetic matter. This possibility is of little consequence for saturation magnetizations obtained at low temperatures, but it must be considered for measurements on nickel at room temperature or higher. It must also be considered in the interpretation of the magnetic changes that occur when vapors are chemisorbed on the sample. This is especially true for nickel as the adsorbent at even moderately elevated temperatures.

The above question is discussed by Bean and Livingston.[19] Some of the evidence tending to establish the existence of such an effect has been obtained on thin films, but Bean has shown that these results are not necessarily applicable to particles such as those under discussion. Particles of cobalt down to about 1.4 nm diameter have been shown by Bean et al.,[22] Cahn et al.,[23] and Knappwost and Illenberger[24] to have normal magnetization–temperature curves. These particles were obtained as precipitates of cobalt from dilute cobalt–copper solid solutions. The only reservation one might have with these conclusions is that the maximum temperature reached in the measurements was only a fraction of the normal Curie temperature for cobalt, namely, 1403 K. On the other hand, Henning and Vogt[25] have reported subnormal Curie points for small particles of iron, and Kneller[26] has found similar results for 2.7 nm diameter particles of Ni_3Mn, $T_C = 743$ K.

As pointed out by Bean and Livingston[19] the determination of T_C in a superparamagnetic sample may be difficult. A reason for this is that the field necessary to produce a significant orientation of the particles may be sufficient to modify the moment of the particle. A possible solution to the problem is, as suggested by Abeledo and Selwood,[27] to rely on Eq. (2.5) but written in the following form:

$$\frac{MM_0}{M_{sp}} = M_0 V \int_0^\infty \left[\coth\left(\frac{vM_0}{k} \cdot \frac{M_{sp}H}{M_0 T}\right) - \frac{kTM_0}{vM_0M_{sp}H} \right] f(v) \, dv \qquad (2.9)$$

One may then plot MM_0/M_{sp} versus $M_{sp}H/M_0T$, using known values of M_{sp}/M_0. This is shown in Fig. 7 for a nickel–silica sample prepared by impregnation of silica gel with nickel nitrate solution, followed by drying and reduction in hydrogen. The particle diameter derived from magnetic measurements (Chapter IV) was $\bar{v} = 3.0$ nm. It will be noted that the measurements were extended to a maximum temperature of 523 K, which is about 100 K below the normal T_C. Superposition of the data is not found

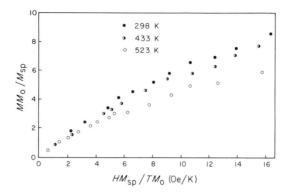

Fig. 7. Magnetization M versus H/T for nickel–silica. The values of M_{sp} are those for a sample of normal T_C.

but, if the M_{sp}/M_0 ratios used are those of a ferromagnetic substance with Curie temperature 565 K instead of the normal 631 K, as in Fig. 8, then normal superposition occurs. Similar results were obtained for particles having diameters up to 8.5 nm.

This method for estimating Curie points in very small particles has been criticized by Binder et al.[28] who, on theoretical grounds, suggest that for such particles no Curie point in the usual sense exists, and that the experimental results are caused by a strong dependence of M_s on temperature. This view is, of course, almost certainly correct as we approach particles containing only a very few atoms. But our conclusion with respect to T_C is that no very large change occurs for nickel down to 3.0 nm diameters and that for temperatures below about 373 K the effect may be ignored.

Still another complication that must be mentioned is the possibility that internal fields between particles of superparamagnetic matter might reveal themselves in a plot of M (or of M/M_s) against T, in the range of low M/T, such that the sample would act as a true paramagnetic. In such a case it should be possible to represent the data in the form of the Curie–Weiss law, $\kappa = C/(T + \Delta)$ in which the Weiss constant is, under certain conditions, a measure of the internal field.

The above condition actually occurs in certain cases but apparently it is of little significance in any of the preparations with which we shall be concerned in later chapters. The matter has been discussed by various authors,[8,20,29,30] but only one related report need be described in any detail. Carter and Sinfelt[31] studied nickel–silica, in a range of particle diameters, above the normal ferromagnetic Curie point for nickel. Under these conditions $\beta(\text{Ni})$ for samples in massive form is, as previously mentioned,

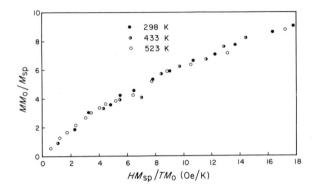

FIG. 8. The data of Fig. 7 replotted for a sample with T_C 70 K lower than the normal, 631 K, for nickel.

considerably higher than the normal value for the saturation moment as obtained from M_0. The experiments permitted a comparison of $\beta(\mathrm{Ni})$ as obtained in the paramagnetic temperature region and also of the Weiss constant (so-called paramagnetic Curie point, Δ) for comparison with T_C. Particle diameters $\bar{v}^2/\bar{v}$ were estimated from the initial slope of M versus H to be described in Chapter IV. Although not specifically stated it appears that the authors used a method[32] for obtaining M_s that has an uncertain physical basis, namely, the "apparent saturation magnetization," obtained well above absolute zero, and stated to be strongly dependent on temperature. We must, therefore, accept the particle sizes given with some reservation although this, in itself, does not necessarily invalidate the conclusions reached.

It was found that as the nickel particle diameter rose from 1.2 nm to virtual infinity, $\beta(\mathrm{Ni})$ rose from 1.1 to the normal value for massive nickel 1.73 (well above T_C) and the Weiss constant, or the paramagnetic Curie temperature ($\theta = -\Delta$), rose from 547 to 638 K. While $-\Delta$ and T_C are not necessarily the same quantities, we may almost certainly agree with the authors that the Δ results show some lowering of the Curie point with diminishing particle size. This is in agreement with conclusions described above for similar samples. With respect to the rising moment with particle size Carter and Sinfelt conclude that the "electronic structure of supported nickel is different from that of bulk nickel." It would be difficult to disagree with that statement or the additional one that the differences observed may be related to differences between surface atoms and bulk atoms in the particle or to the suggestion that the catalyst support influences the electronic structure of the nickel. On the other hand, we shall present evidence in Chapter V that M_0 for such particles is within $\pm 1\%$

of the value for massive nickel. Perhaps the electronic arrangement in nickel at 4 K is different from that at 700 K. (Every other physical property is different.) But we shall not attempt to answer that question here.

References

1. A. Michel, *Ann. Chim. (Paris)* **8**, 317 (1937).
2. A. Michel, R. Bernier, and G. Le Clerc, *J. Chim. Phys.* **47**, 269 (1950).
3. J. Frenkel and J. Dorfman, *Nature (London)* **126**, 274 (1930).
4. W. C. Elmore, *Phys. Rev.* **54**, 1092 (1938).
5. L. Néel, *Ann. Geophys.* **5**, 99 (1949).
6. E. C. Stoner, *Phil. Trans. Roy. Soc. London Ser. A* **235**, 165 (1936).
7. R. Gans and P. Debye, *in* "Handbuch der Radiologie" (E. Marx, ed.), Vol. VI, p. 719. Leipzig, 1925.
8. C. P. Bean and J. D. Livingston, *J. Appl. Phys.* **30**, 120S (1959).
9. I. S. Jacobs and C. P. Bean, *in* "Magnetism" (G. T. Rado and H. Suhl, eds.), Vol. 3, p. 271. Academic Press, New York, 1963.
10. R. M. Bozorth, "Ferromagnetism," pp. 567–568. Van Nostrand, New York, 1951.
11. C. Kittel, "Introduction to Solid State Physics," 2nd ed., p. 429. Wiley, New York 1960.
12. J. J. Becker, *Trans. AIME* **209**, 59 (1957).
13. R. E. Dietz and P. W. Selwood, *J. Chem. Phys.* **35**, 270 (1961).
14. J. J. B. van Eijk van Voorthuysen and P. Franzen, *Rec. Trav. Chim.* **70**, 793 (1959).
15. R. E. Watson, P. Fulde, and A. Luther, *AIP Conf. Proc. Magnetism and Magnetic Materials* **10**, (2), 1535 (1973).
16. L. Liebermann and J. Clinton, *AIP Conf. Proc. Magnetism and Magnetic Materials* **10**, (2), 1531 (1973).
17. C. A. Neugebauer, *Phys. Rev.* **116**, 1441 (1959).
18. J. C. Walker, C. R. Guarnieri, and R. Semper, *AIP Conf. Proc. Magnetism and Magnetic Materials* **10**, (2), 1539 (1973).
19. C. P. Bean and J. D. Livingston, *J. Appl. Phys.* **30**, 126S (1959).
20. S. Takajo, S-i. Kobayashi, and W. Sasaki, *J. Phys. Soc. Jap.* **35**, 712 (1973).
21. T. Shinjo, T. Matsuzawa, and T. Takada, *J. Phys. Soc. Jap.* **35**, 1032 (1973).
22. C. P. Bean, J. D. Livingston, and D. S. Rodbell, *J. Phys. Radium* **20**, 298 (1959).
23. J. W. Cahn, I. S. Jacobs, and P. E. Lawrence, quoted by Bean and Livingston in Ref. 8.
24. A. Knappwost and A. Illenberger, *Naturwissenschaften* **45**, 238 (1958).
25. W. Henning and E. Vogt, *J. Phys. Radium* **20**, 277 (1959).
26. E. Kneller, *Z. Phys.* **152**, 574 (1958).
27. C. R. Abeledo and P. W. Selwood, *J. Appl. Phys.* **32**, 229S (1961).
28. K. Binder, H. Rauch, and V. Wilderspaner, *J. Phys. Chem. Solids* **31**, 391 (1970)
29. W. F. Brown, *J. Appl. Phys.* **30**, 130S (1959).
30. P. O. Voznyuk and V. N. Dubinin, *Ukr. Fiz. Zh.* **19**, 160 (1974).
31. J. L. Carter and J. H. Sinfelt, *J. Catal.* **10**, 134 (1968).
32. J. L. Carter, J. A. Cusumano, and J. H. Sinfelt, *J. Phys. Chem.* **70**, 2257 (1966).

III

Magnetization Measurements at High M/M_0

1. The Experimental Problem

Our chief purposes are to show how magnetic methods may be used to measure particle size and to gain information concerning the binding of an adsorbate to an adsorbent. One method of achieving the latter goal is, in appropriate systems, to measure the change in average magnetic moment of adsorbent atoms per molecule of vapor adsorbed. This chapter will be devoted to the experimental arrangements for making such measurements.

For true paramagnetic matter it suffices to measure the susceptibility over a moderate range of temperature so that one may calculate the Bohr magneton number as shown in Eq. (1.6). This is possible because in paramagnetics all the (atomic) particles have the same magnetic moment. But in superparamagnetic matter the particle size is rarely, if ever, uniform and hence the moment of a particle, $m_p = M_{sp}v$, is far from uniform. If one deals with, say, nickel that is superparamagnetic, or ferromagnetic, accurate determination of β(Ni) requires measurements of M_s at temperatures sufficiently low that extrapolation to find M_0 is feasible. Calculation as shown on p. 11 then gives the desired magneton number. These remarks should not be construed as meaning that magnetic measurements at relatively low values of H/T are of no value in the study of supported nickel and cobalt, but accurate estimates of the change in β(Ni) or β(Co) produced by a chemisorbed molecule cannot be made in that way.

Our purposes, therefore, require that magnetization should be measured at fields high enough and at temperature low enough so that extrapolations to find M_0 may be made with confidence; this means fields well in excess of 10 kOe and preferably much higher should be used. It also means the temperature of liquid helium and preferably considerably lower should be used. The samples to be studied are pyrophoric in air and hence must be reduced and subsequently handled in closed containers. We shall require a quantitative determination of the amount of adsorbent present. If this adsorbent is a metal it will probably be necessary to obtain the amount actually present as reduced metal and not as oxide or other nonadsorbing form. It will also be necessary to make a quantitative measurement of the amount of vapor adsorbed. Some complications are introduced by the necessity that during measurements of magnetization the sample must be at such low temperature. Other requirements are that the sample should be in a uniform field during measurement, and that provision should be made for carrying out chemical treatment such as reduction in flowing hydrogen at near 700 K, evacuation, and so forth, all *in situ*.

2.   The Weiss Extraction Method[1]

The following description is based on the adaptation developed by Dietz[2] for the study of nickel–silica preparations, and for the handling and measurement of gases in contact with the sample.

The field is produced by a 12-inch electromagnet with pole face 4 inches and gap 2.75 inches. The field is variable from 0 to about 18 kOe, and reversible. Within a volume of about 1 in.³ between the pole tips field uniformity was within a small fraction of 1%. The field was monitored with a

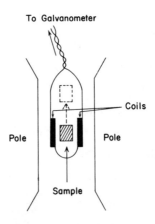

FIG. 9.   Relation of sample to sensing coils for measurement of M_s.

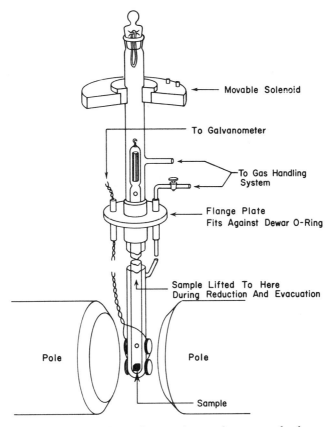

FIG. 10. The arrangement of sample, container, and magnet poles for measurement of M_s.

commercial meter.[3] The magnet was mounted on tracks to permit lateral displacement. Considerably higher fields, with corresponding improvement in precision, may be achieved by the use of a superconductive magnet,[4] up to 100 kOe.[5]

The sample consists of one pressed cylindrical pellet containing about 1 g of metal. This is placed in the geometric center of the pole gap. Two small Helmholtz sensing coils are placed coaxially to the pole pieces—one on each side of the sample. These coils are about 1 cm in diameter and consist of 1000 turns of No. 40 ceramic-insulated copper wire. The coils are connected in series to a ballistic galvanometer or other integrating device.

When a measurement is to be made the sample is lifted mechanically to a position a short distance above the coils, but still in the region of maximum magnetic field. In this way the lines of force passing through the sample are forced to cut the coils in such a manner as to induce a current in the coils. This current is integrated by the ballistic galvanometer to give a reading which may be shown[3] to be proportional to the magnetization of the sample. The arrangement is shown diagrammatically in Fig. 9. The method of raising, or lowering, the sample between the sensing coils is shown in Fig. 10.

If the magnet power supply is adequately stabilized the arrangement described above may give sufficient precision. Otherwise it may be necessary to provide two pairs of sensing coils connected as shown in Fig. 11. The sample is then raised from between one pair of coils to a position between the second pair. This arrangement balances out transient changes in the applied field. All the coils are in series, but the upper pair of coils is wound in opposition to the lower.

Measurements in the liquid-helium region are made with the sample surrounded by a Dewar flask of conventional design for this purpose except that the portion of the Dewar between the magnet poles is shielded by a copper screen cooled by, and projecting down from, the liquid nitrogen shield. Measurements at the λ point of helium, 2.18 K, are made by pumping on the liquid helium in the usual way. The samples are, of necessity, handled in the absence of air after reduction. They are conveniently reduced *in situ* by raising the sample holder to a position high enough to

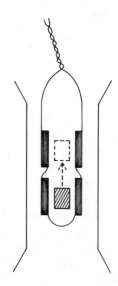

FIG. 11. Two pairs of sensing coils to minimize the effect of field fluctuations during measurement of M_s.

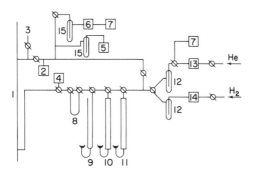

FIG. 12. Gas handling system for saturation magnetization studies: (1) apparatus shown in Fig. 11, (2) cold cathode gauge, (3) hydrogen exhaust, (4) turntable McLeod gauge, (5) McLeod gauge, (6) oil diffusion pump, (7) mechanical pumps, (8) closed-arm oil manometer, (9) open-arm mercury manometer, (10) gas microburet, (11) gas buret, (12) silica gel traps, (13) helium purification train, (14) hydrogen purifier, (15) traps cooled with liquid nitrogen.

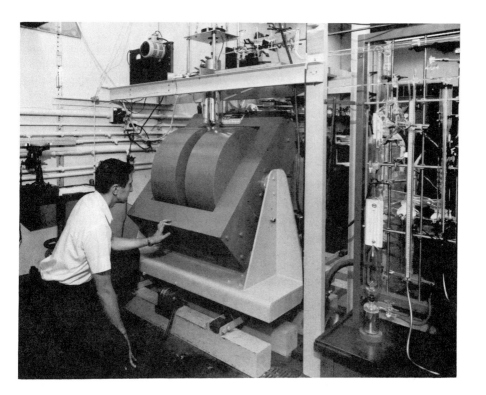

FIG. 13. Magnet assembly for measurement of M_s at low temperatures.

avoid damage to the sensing coils. A small sleeve furnace provides the proper reduction temperature.

The gas handling system is also of conventional design as shown in Fig. 12. In brief, the sample is reduced by flowing purified hydrogen for many hours. For a typical nickel–silica sample the reduction temperature is 633 K. The sample is then evacuated for a minimum of 2 hours at 633 K to a pressure of about 10^{-4} N·m^{-2}, and allowed to cool to the temperature of measurement. A trace of purified helium is added to promote attainment of thermal equilibrium. After the magnetization is measured at whatever temperature is desired, the sample is warmed to room temperature, after which a measured volume of adsorbate gas is admitted. The pressure in the dead space must, of course, be kept low. The sample is then cooled again for a final measurement of magnetization as affected by a known quantity of adsorbed vapor. Some idea of the complete assembly may be gained from Fig. 13.

3.   Correction for Demagnetizing Field

The field acting to magnetize a sample is always less than the field in the absence of the sample. The reason for this is that the free poles at the ends of the oriented dipoles produce a demagnetizing effect which is dependent on the shape of the sample and on its magnetization. For samples of the kind under consideration the demagnetization correction has been considered by Trzebiatowski and Romanowski[6] and by Dietz.[2]

The actual field H is related to the apparent applied field H_{app} by the expression

$$H = H_{\mathrm{app}} - \eta M_{\mathrm{T}} \qquad (3.1)$$

where η is the demagnetization constant, and M_{T} is the magnetization of the whole sample. Demagnetization constants have been calculated for samples of various shapes; for spheres, $\eta = \frac{4}{3}\pi$. The samples of silica-supported metals used in this kind of investigation are short cylinders (pellets), but we know little concerning the shape of the metal particles within each pellet. We do, however, know the saturation magnetization M_{s} for several metals; and the volume fraction V/V_{T} of ferromagnetic substance present is readily found, V_{T} being the total volume of a sample, including the silica (or other) supporting medium. If we write

$$\eta M_{\mathrm{T}} = \eta \cdot \frac{M}{M_{\mathrm{s}}} \cdot M_{\mathrm{sp}} \cdot \frac{V}{V_{\mathrm{T}}} \qquad (3.2)$$

then, for instance, from Fig. 14 it is seen that at room temperature and 5

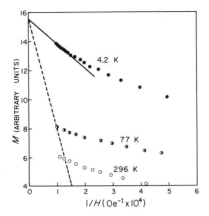

FIG. 14. Magnetization versus $1/H$ for a reduced nickel–silica at several temperatures. This shows how important it is to obtain data at low temperatures if a meaningful extrapolation to M_s at 0 K is required.

kOe the fraction $M/M_s = 0.34$, thus as $V/V_T \simeq 0.1$, ηM_T must be about 114 Oe.

One sees, therefore, that the demagnetizing field will not be a negligible fraction of H_{app} until saturation is approached at fields of the order of 10^4 Oe. Figure 15 shows data on a sample of massive nickel obtained at 77 K before and after correction for demagnetization. As expected, the curves converge at high field.

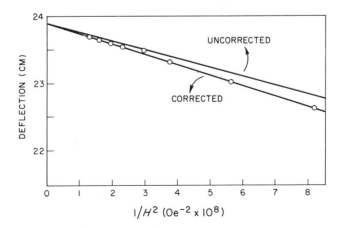

FIG. 15. The approach to M_s for massive nickel before and after correction for demagnetizing fields.

4.   Correction for the Magnetic Image Effect

A specimen between the poles of a magnet may induce magnetic charges, or images, in the pole tips. These images have the effect of increasing the apparent magnetization of the sample, and the result can be serious. This effect was studied by Weiss and Forrer,[1] and for catalysts by Dietz.[2]

It is reasonable to assume that at very large pole gaps the image effect will be negligible. The magnitude of the effect may be demonstrated by measuring the apparent magnetization of a sample at constant field. Results on a coprecipitated nickel–silica sample are shown in Fig. 16 for several fields. It is assumed that the true magnetization is being measured when the apparent magnetization becomes independent of the pole gap.

The apparent magnetization M_{app} is thus equal to the true magnetization M plus an added contribution caused by the image effect. We may write

$$M_{\mathrm{app}} = M[1 + f(w, \mu)] \tag{3.3}$$

where $f(w, \mu)$ is a function of the pole gap and of the permeability of the pole tips. Then if M_{app} is measured at two different field strengths H_1

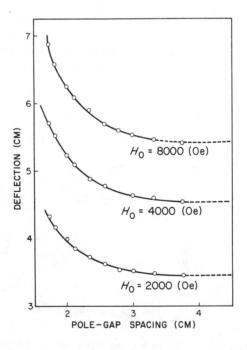

$H_0 = 8000$ (Oe)

$H_0 = 4000$ (Oe)

$H_0 = 2000$ (Oe)

DEFLECTION (CM)

POLE–GAP SPACING (CM)

FIG. 16.   Apparent M at several pole gap spacings.

and H_2:

$$\frac{M_{\mathrm{app}}(H_1)}{M_{\mathrm{app}}(H_2)} = \frac{M(H_1)[1 + f(w, \mu_1)]}{M(H_2)[1 + f(w, \mu_2)]} \tag{3.4}$$

and

$$\frac{M_{\mathrm{app}}(H_1)}{M_{\mathrm{app}}(H_2)} = \frac{M(H_1)}{M(H_2)} \tag{3.5}$$

provided $f(\mu_1) \simeq f(\mu_2)$. That this view is correct may be shown by normalizing the values of apparent magnetizations shown in Fig. 16 by the magnetization corresponding to the same pole gap. When this is done it is found that the normalized values are independent of gap over the range 1–8 kOe investigated.

The correction for the image effect may now be found in the following manner. All specimens investigated show some remanence, although for some samples this does not become measurable until we reach quite low temperatures. We measure $M_{r'}$ at a given pole gap, but with the field equal to zero as measured by the gaussmeter. The magnet is then moved out of position so that the pole gap is essentially infinite, and the true remanence M_r is measured. The ability to make this kind of measurement is a virtually decisive argument in favor of the Weiss or similar experimental methods. However, there is some possibility that an error can arise from the change of μ with H.

In a typical case the data found at a gap of 2.74 cm, and given in centimeters of deflection on the ballistic galvanometer are as follows: $M_r/M_{r'} = 2.66/2.76 = 0.964$. We may then find corrected values of M from the observed values of M_{app} by multiplying by 0.964. A different correction must, of course, be used if the pole gap is changed.

5. Calibration

The chief objective of this work is to measure the change of magnetization produced by a molecule of adsorbate. This does not require anything more than a ratio of magnetizations before and after adsorption. But for the interpretation of these data in terms of electronic interaction we must know the absolute magnetizations. In principle these may be found from a consideration of the geometry and constants of the experiment, but a much simpler procedure is to use as calibrating agent a sample the magnetization of which is precisely known. Pure nickel is especially suitable for measurements on nickel catalyst samples.

A sample of powdered, polycrystalline nickel is mixed with silica gel to prepare a pellet similar in size and volume concentration of nickel to those used for the adsorption studies. The sample is heated, *in situ*, in hydrogen for 12 hr at 623 K. While this step may be thought to be scarcely necessary, it may result in a small but definite increase of magnetization, suggesting some superficial oxidation in the sample as obtained.

At high fields the approach to saturation of a ferromagnetic is described by a $1/H^2$ law, as follows:

$$M = M_s(1 - b/H^2) \tag{3.6}$$

where b is a constant. This was found to hold in the present case. One need, therefore, only measure the galvanometer deflections for a given mass of nickel at several fields, extrapolate to $1/H = 0$,* and equate the deflection so found to the known value of M_s at the temperature of calibration.

6.   The Faraday Method[7]

Of the various classical methods for measuring magnetic susceptibility that of Faraday lends itself to the study of adsorption and reaction processes by samples of interest in heterogeneous catalysis. The principle involved is that a sample placed in a nonhomogeneous field suffers a displacing force. If the sample has a positive susceptibility the displacement will be in the direction of increasing field intensity. A very simple adaptation of the Faraday method to an adsorption study is shown in Fig. 17. The field gradient in this case is vertical and the field decreases in the upward direction because of the particular shape of the pole faces. The force on the sample is

$$\text{force} = \kappa V H \delta H/\delta s = M V \delta H/\delta s \tag{3.7}$$

where κ is the susceptibility, V the volume, H the average field strength, and $\delta H/\delta s$ the field gradient along the vertical axis. The product $\kappa V = m\kappa/\rho$, where m is the mass of the sample and ρ the density. The Faraday method does not permit the sample to be placed in a uniform field, but this is rarely a serious problem. The change of field is almost always small over the volume of the sample.

Various authors[8] have used the Faraday method of a design appropriate for our purposes. This description will be confined to the adjustable field gradient magnetometer adaptation of Lewis.[9,10] In this system the sample hangs vertically as shown in Fig. 17 but is suspended from a recording electric vacuum microbalance, and the method for producing the necessary field gradient is quite different from that indicated above. Instead of the

* Or to $1/H^2 = 0$. See C. P. Bean and I. S. Jacobs, *J. Appl. Phys.* **31**, 1228 (1960).

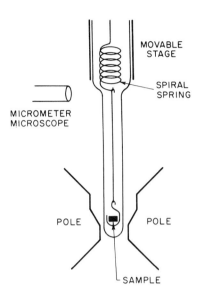

FIG. 17. Faraday balance using vertical suspension from a spiral spring.

usual specially shaped pole gap there are pairs of coils in a "figure 8" arrangement. These are mounted on the faces of parallel poles. With an appropriate direction of current through these coils the field produced from, say, the upper coil opposes that of the large magnet, and that from the lower coil adds to it. The currents necessary to produce acceptable field gradients are fairly large, and thus proper cooling must be maintained, but this may be done without serious impairment of the space necessary for the sample and its temperature and atmosphere control.

The use of field gradient coils offers several advantages. The field (as opposed to the gradient) may be altered at will up to at least 20 kOe. The gradient may be changed from about $+0.5$ to -0.5 kOe over the sample height, and this may be done very rapidly and independently of the main field. The field, and the gradient, may be controlled and recorded electronically. Together with electronic measurement of the sample mass in zero gradient, the method offers a range of flexibility and convenience not approached by the older methods.

The method lends itself readily to pretreatment of samples *in situ* over a very wide temperature range. With respect to the quantitative simultaneous measurement of adsorbed gases it is clear that the dead volume is too large for this to be done volumetrically. But for a wide range of adsorbates such measurements are possible gravimetrically without much loss of precision.

Other methods have been described for measuring magnetizations. One of these is the vibrating magnetometer,[11] but this method does not appear to have received extensive application to problems in chemisorption and related areas. The basic magnetic theory of all these methods is given by Zijlstra.[12]

7. General Procedure

The chief requirements for adsorption studies are that the particles should be small enough so that an appreciable quantity of adsorbate may be taken up, yet not so small that difficulty is encountered in extrapolating the measured magnetizations to obtain a reasonably accurate value of M_0. It is not a requirement for the success of saturation studies that the sample should exhibit superparamagnetism. In view of this, preparation procedures will vary widely, depending on the particular adsorbent under study, and the problem at hand. Commercially available nickel–kieselguhr catalysts containing 30–50% nickel are generally suitable. Supported cobalt may be made by impregnation of high-area silica gel with cobaltous nitrate solution, followed by drying, careful ignition, and reduction. Experienced workers in the field of heterogeneous catalysis will be able to think of various other preparative procedures.

Analysis for the total quantity of metal present, either as metal or combined, will offer no difficulty, but accurate determination of the fraction of reduced metal may prove troublesome. In the case of nickel there are several analytical procedures which may be tried, and compared. One method is to measure the volume of hydrogen taken up during reduction. This may be done by circulating a measured volume of hydrogen over the weighed sample in a closed system.[13,14] The water formed is frozen in a trap. This procedure is reasonably satisfactory except that the catalyst support may contain an appreciable amount of residual water which is slowly released at the temperature of reduction. It must also be remembered that some hydrogen will be chemisorbed on the metal as it is formed. For very highly dispersed nickel this may amount to 20% of the whole volume of hydrogen used.

A related method often used for nickel is to place the weighed sample, after reduction, in hydrochloric acid. The displaced hydrogen is collected and measured.

Another chemical method[15,16] applicable to nickel is based on the von Wartenburg reaction of sulfur vapor with nickel oxide to form sulfur dioxide and nickel sulfide; with nickel metal the reaction is simply the formation of the sulfide. The sulfur dioxide may be determined iodimetrically, or it may be oxidized in hydrogen peroxide to sulfuric acid, which is then titrated. This method appears to be the most nearly reliable for our purpose, although Eggertsen and Roberts[15] express little confidence in the method as applied to nickel supported on alumina. Some difficulty may be experienced because at the temperature of reaction, namely, 1123 K, some residual water from a silica catalyst support may react with the sulfur to form sulfur dioxide and hydrogen sulfide. This may be detected by the formation of colloidal sulfur in the effluent. A method for combatting this difficulty is to preheat the sample to 923 K in an inert atmosphere, but this may cause some changes in the proportion of metal present.

Still another[16] chemical method involves reaction of the reduced metal with a bromine–methanol solution. Apparently the method is not appropriate for all preparations of supported nickel.

Under certain circumstances a determination of M_0 may be expected to yield a satisfactory estimate of the fraction of reduced metal present. Knowing the total mass of metal present in both reduced and oxidized form we may readily calculate M_0 for 100% reduction. Then $M_0(\text{obs})/M_0$ (expt) gives the fraction reduced. This procedure may well be valid if we may safely assume that M_0 for very small particles is the same as for massive metal. This question was discussed on p. 24. A few authors have attempted to substitute $M_0(\text{obs})$ by M_s estimated by extrapolation from measurements at or near room temperature. At best this can give only a rough estimate of M_0. Others have tried to improve the accuracy by sintering the sample in an inert atmosphere at a temperature high enough to increase the metal particle size. This makes a more accurate estimate of M_s possible but it introduces other uncertainties.

This section will be concluded with some remarks concerning precision. The absolute saturation magnetizations approach a precision of $\pm 1\%$, or better under favorable conditions. But the relative magnetizations before and after vapor adsorption are accurate to about $\pm 0.1\%$. A typical nickel–silica sample is capable of chemisorbing nearly 20 cm³ of hydrogen per gram of nickel at room temperature. Complete coverage cannot be utilized because of the necessity for keeping the quantity of gas in the dead space negligible. The dead space, as closed off during actual measurement, may be about 92 cm³. The volume (STP) of hydrogen taken up in a typical experiment is of the order of 8 cm³·g⁻¹ of nickel, and this may be measured with at least the same precision as the magnetization. Over all precision in determining the change of magnetization per cubic centimeter of gas absorbed is thus rather better than $\pm 1\%$ in favorable systems, less in others.

References

1. P. Weiss and R. Forrer, *Ann. Phys. (Paris)* **5**, 153 (1926).
2. R. E. Dietz and P. W. Selwood, *J. Chem. Phys.* **35**, 270 (1961).
3. M. B. Stout, "Basic Electrical Measurements," pp. 370–373. Prentice-Hall, Englewood Cliffs, New Jersey, 1950.
4. G. A. Martin and P. Fouilloux, *J. Catal.* **38**, 231 (1975).
5. J. T. Richardson, personal communication.
6. W. Trzebiatowski and W. Romanowski, *Rocz. Chem.* **31**, 1123 (1957).
7. M. Faraday, "Experimental Researches," Vol. III, pp. 27, 497. Taylor and Francis, London, 1855.
8. J. T. Richardson and J. O. Beauxis, *Rev. Sci. Instrum.* **34**, 877 (1963).
9. R. T. Lewis, *Rev. Sci. Instrum.* **42**, 31 (1971).
10. R. T. Lewis, *J. Vacuum Technol.* **11**, 404 (1974).
11. D. J. Craik, D. D. Eley, and R. J. Mellar, *Trans. Faraday Soc.* **65**, 1649 (1969).
12. H. Zijlstra, "Experimental Methods in Magnetism (2)." North-Holland Publ. Co., Amsterdam, Wiley, New York, 1967.
13. F. N. Hill and P. W. Selwood, *J. Am. Chem. Soc.* **71**, 2522 (1949).
14. V. C. F. Holm and A. Clark, *J. Catal.* **11**, 305 (1968).
15. F. T. Eggertsen and R. M. Roberts, *Anal. Chem.* **22**, 924 (1950).
16. G. A. Martin, B. Imelik, and M. Prettre, *J. Chim. Phys.* **66**, 1682 (1969).

IV

Magnetic Particle Size
Determination

1. Granulometry

There is increasing evidence that particle size and geometry have significance in catalyst activity and specificity. There are essentially two methods, each in several variations, by which magnetic measurements will yield particle size information. Some of these measurements may be made under conditions of actual catalytic reactivity, without the necessity of sample removal. All of the methods are restricted to ferromagnetic, or superparamagnetic, matter. Certain aspects of the area have been surveyed by Whyte,[1] and reviewed by Spindler.[2]

Equation (2.2) shows that the particle volume v appears in two terms in the complete expression for an assembly of superparamagnetic particles. For measurements at low M/M_s the second term may be neglected and analysis (to be given later) yields $\bar{v}^2/\bar{v}$. This procedure for obtaining $\bar{v}$ will be referred to as the Langevin low-field method (LLF). But for measurements under conditions such that $M \simeq M_s$, analysis yields $\bar{v}$ directly. This will be referred to as the Langevin high-field method (LHF). Reference to Eq. (2.8) shows that v is also obtainable from relaxation time t measurements. A procedure based on this will be referred to as the Néel relaxation (NR) method.

Several groups have published particle size determinations on small particles of nickel, cobalt, and iron and have compared the results obtained by one or more magnetic methods with those found by other methods. The other, nonmagnetic, methods include electron microscopy, x-ray line broadening, small angle x-ray scattering, and chemisorption. All workers in this area know that when there is a particle size distribution, the sizes obtained by different methods are not directly comparable. Pulvermacher and Ruckenstein[3] have provided tables permitting a comparison of the different averages to be made. Thus the average value for v, obtained by each of the several methods, is as follows:

magnetic (LLF)	$f_0{}^\infty v^2 n(v, t)\ dv / f_0{}^\infty vn(v, t)\ dv$
magnetic (LHF)	$f_0{}^\infty vn(v, t)\ dv / f_0{}^\infty n(v)\ dv$
magnetic (NR)	any average needed
chemisorption	$\{f_0{}^\infty vn(v, t)\ dv\}^3 / \{f_0{}^\infty v^{2/3} n(v, t)\ dv\}^3$
x-ray line broadening	$\{f_0{}^\infty v^{4/3} n(v, t)\ dv\}^3 / \{f_0{}^\infty vn(v, t)\ dv\}^3$
small angle scattering	$\{f_0{}^\infty v^{7/3} n(v, t)\ dv\}^{3/2} / \{f_0{}^\infty v^{5/3} n(v, t)\ dv\}^{3/2}$
electron microscopy	any average needed

In these relationships $n(v, t)\ dv$ is the number of particles per unit area of support having a volume in the range v to $(v + dv)$. Of the several methods, LHF and NR yield a radius independent of the form of size distribution.

It is essential for all the magnetic particle size methods that the sample be free of chemisorbed molecules before the measurements are made. In later chapters it will be shown how the magnetization decreases as molecules of hydrogen (and other substances) are chemisorbed on the surface, and it might be thought that the maximum fractional decrease would be a measure of the number of surface metal atoms present. This is true if the saturation magnetizations are measured but this, in turn, presents experimental difficulties. The method has been explored by Carter et al.[4] who point out the problems involved. Nevertheless, an indirect estimate of M_s yields a fraction of nickel on the surface in reasonably satisfactory agreement with that obtained by other methods.

2. The Langevin Low-Field (LLF) and High-Field (LHF) Methods[2,5,6]

From Eq. (2.2), for uniform particles at low fields and elevated temperature

$$M = M_{sp} V M_{sp} v H / 3kT \tag{4.1}$$

where v is the volume of a particle and $V = N_p v$ where N_p is the number of particles in a sample. Then

$$M = \frac{M_{\mathrm{sp}} H}{3kT} \sum N_{\mathrm{p}} v^2 \qquad (4.2)$$

Also, for all the measurements to be described, M_{s} is approximately equal to $M_{\mathrm{sp}} \sum N_{\mathrm{p}} v$; hence, it is convenient to use relative magnetizations

$$\frac{M}{M_{\mathrm{s}}} = \frac{M_{\mathrm{sp}} H \sum N_{\mathrm{p}} v^2}{3kT \sum N_{\mathrm{p}} v} \qquad (4.3)$$

and the average particle volume $\bar{v}^2/\bar{v}$ may be obtained from the initial slope (i.e., the slope at low field) of the curve of M plotted with respect to H/T, as follows:

$$\frac{\bar{v}^2}{\bar{v}} = \frac{3kT}{M_{\mathrm{sp}} H} \left(\frac{M}{M_{\mathrm{s}}} \right) \qquad (4.4)$$

From Eq. (4.4) it is obviously necessary to obtain M_{s} before $\bar{v}$ may be calculated. Figure 14 shows how important it is to use high field and low temperature to estimate M_{s} with any degree of accuracy on samples of this kind. Measurements are, therefore, first made on the sample at the boiling point of helium or, preferably, even lower. A plot of magnetization (or of galvanometer deflection) versus reciprocal field will then give a value corresponding to M_{s}. Thus, in Fig. 14, M_{s} is 15.6 (in arbitrary units). The sample is then warmed to conditions under which true superparamagnetism is exhibited as evidenced by M versus H/T superposition. For the sample for which data are shown in Fig. 14 this occurs at, and above, 77 K. The initial slope of the magnetization as a function of field is found from Fig. 4. Taking, say, the slope of the initial points at 296 K, we find a deflection of 3.1 at 10^3 Oe. This gives $M/M_{\mathrm{s}} = 3.1/15.6 = 0.199$. Then, assuming that the spontaneous magnetization of very small particles of nickel is the same as that of massive nickel, namely, 485 Oe at 296 K, we have

$$\bar{v}^2/\bar{v} = \frac{3 \times 1.38 \times 10^{-6}\ \mathrm{erg \cdot K^{-1}} \times 2.96 \times 10^2\ \mathrm{K} \times 1.99 \times 10^{-1}}{4.85 \times 10^2\ \mathrm{Oe} \times 10^3\ \mathrm{Oe}}$$

$$= 50 \times 10^{-21}\ \mathrm{cm^3}$$

We now consider the high-field approximation. For $M/M_{\mathrm{s}} \simeq 1$, the Langevin function becomes

$$\frac{M}{M_{\mathrm{s}}} = 1 - \frac{kT}{m_{\mathrm{p}} H} \qquad (4.5)$$

which, for the situation under consideration, becomes

$$\frac{M}{M_s} = 1 - \frac{kT \sum N_p}{M_{sp}H \sum N_p v}$$ (4.6)

and this yields

$$\bar{v} = \frac{kT}{M_{sp}H} \cdot \frac{1}{1 - M/M_s}$$ (4.7)

In spite of the advantages of the high-field approach it must be pointed out that it is at high field (and low temperature) than the effects of anisotropy become important. A high-field estimate[5] on the same sample as used for the low-field value gave $\bar{v} = 4.5 \times 10^{-21}$ cm^3. These results are about as satisfactory as could be expected in view of the fact that the average $\bar{v}$ must always be smaller than $\bar{v}^2/\bar{v}$. Experimentally, the Weiss extraction method and the Lewis adaptation of the gradient coil Faraday method, as described in Chapter III, are suitable for LLF and LHF determinations.

There have been several attempts to use Eqs. (4.5) and (4.7) without the rather tedious direct determination of M_s at low temperatures Heukelom et al.[7] found that M as obtained over a range of field strength at room temperature could be extrapolated from the empirical relation

$$\frac{1}{M} = \frac{1}{M_s} + \frac{1}{M_s(aH)^{0.9}}$$ (4.8)

where a is a constant. The method, as further developed by Trzebiatowski and Romanowski[8] and Trzebiatowski,[9] is illustrated in Fig. 18 from data obtained by Dietz and Selwood[6] who also measured M_0 directly on the same sample. It was found that M_s as found by the Heukelom–Trzebiatowski method is about 9% lower than that of pure massive nickel at the same temperature. It appears, therefore, that there is no real substitute for determinations at high field and low temperature, although it is doubtful if any particle size determination by magnetic methods is more accurate than ±20%. Some examples of results obtained by this method are given by Richardson.[10]

Within the limits of superparamagnetic behavior the plot of M versus H/T, as previously mentioned, owes its slope at lower values of M/M_s to the larger particles in a sample, and at higher M/M_s to the smaller, less readily magnetized, particles. It is, therefore, possible to obtain particle size distribution curves by solving for the number of particles of a given volume from the appropriate set of linear equations, or by the method of trial and error. This procedure has been used by Romanowski et al.[11] and Romanowski[12] to obtain size distributions for supported nickel and cobalt in several

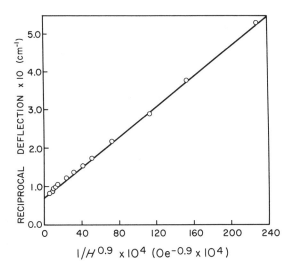

$1/H^{0.9} \times 10^4 \ (Oe^{-0.9} \times 10^4)$

FIG. 18. A plot M versus $1/H^{0.9}$, from Dietz, and his value of the true M_s at 0 K, namely, M_0. This permits a comparison of M_0 obtained by the Heukelom–Trzebiatowski extrapolation with that actually measured at high H and low T. (Dietz and Selwood, Ref. 6.)

preparations. For instance, it was found that a coprecipitated nickel magnesium carbonate, after reduction at 623 K had 7% of particles too large to show superparamagnetism and, for the remainder, a size radius maximum at about 1.0 nm. Reduction at 673 K, as expected, shifted the maximum to a larger radius. Similar results were obtained for supported cobalt except that, presumably owing to the larger anisotropy constant, a larger fraction of the particles may be detected at lower radius. Romanowski points out that the method used to determine M_s (heating the samples for a long time at temperature considerably higher than necessary for reduction) is not always reliable. This is especially true for polydisperse systems.

A more general theoretical treatment of the method used by Romanowski et al. is given by Dreyer,[13] and some refinements in interpretation, especially with respect to anisotropy effects, are described by Schwarz.[14] In addition, extensive theoretical and experimental studies of particle size distributions obtained by the Langevin, and other, methods, have been made by Charcosset et al.[15] and by Martin et al.[16–20] The more important conclusions from these papers will be presented in a later section.

3. The Néel Remanence (NR) Method

Reference to Eq. (2.8) shows that a particle size determination method may be based on the relaxation phenomenon described by Néel (see p.

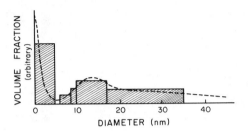

F IG. 19. Distribution of particle volumes in a Raney nickel sample as obtained by the (NR) Néel relaxation method (after Weil, Ref. 21).

20). An application of this method to a system of interest in heterogeneous catalysis, Raney nickel, was made by Weil.[21] (Raney nickel is an almost impossibly complicated system on which to make measurements of this kind.) The procedure used was the Weiss extraction method with measurement of the decay rate of the remanent magnetization M_r after removal of the sample from the field. Then, from tables of critical decay rates (see p. 21) and measurements at various temperatures, it is possible to construct a particle size distribution curve for the sample as shown in Fig. 19.

There have been numerous studies of supported nickel by the use of an ac permeameter. This convenient apparatus (described in Chapter VI) is limited in its applicability by the necessity that the sample should exhibit superparamagnetic behavior. In other words, the relaxation time of the metal particles must be short compared with the duration of the applied power cycle. But Martin[16] has taken advantage of this very same difficulty by making measurements of M over a range of both temperature and frequency. In this way estimates of particle volumes have been obtained down to quite small sizes, and with considerable convenience. In the use of this method consideration must be given to the diminution of M_{sp} as the Curie point is approached, and also to the effect of interparticle interaction on the relaxation time.[22]

4.   Comparisons of Methods

In this section we present examples of particle size determinations. For each of these examples the authors have provided data obtained by one or more other methods on the same, or very similar, samples. An early example of this kind of information is given by Dietz and Selwood[6] who found that a nickel–silica preparation gave an (LLF) particle diameter of 5.0 nm, an (LHF) diameter of 2.0 nm, and a hydrogen chemisorption diameter of 8.1 nm (corrected for arithmetical error in the original).

Charcosset *et al.*[15] have reported on two nickel preparations, the one formed by impregnation of silica–alumina with nickel formate solution followed by decomposition to metal at 543 K, the other by coprecipitation of the nickel from solution followed by reduction in hydrogen at 623 K. Total nickel was determined chemically. The fraction of nickel present as metal was determined essentially by extrapolation to find M_s from the magnetization curves obtained at 77, 190, and 295 K. Agreement with chemical determination of metal present was satisfactory for the hydrogen-reduced sample. For the impregnation samples the fraction of total nickel found to be present as metal was much less than 100%, and this presented a problem at low temperature and high field. Unreduced Ni^{2+} ions make a negligible contribution to the magnetization at, or above, room temperature, but it has been suggested that they might make a measurable contribution at very low temperature. This is certainly true, but not at the lowest temperature used in this work. The authors used the chemisorption of oxygen to obtain the accessible metal surface in each sample. This procedure is acceptable if carefully done to prevent oxidation below the metal surface. If there is any obvious criticism of the work, in a difficult area, it is that in an effort to prevent air from reaching the nickel prior to the magnetic measurements the sample was covered with a benzene solution that was then supposedly removed by gentle heating under reduced pressure. But it is very doubtful if this would remove chemisorbed benzene which, like nearly all adsorbates, quite appreciably lowers the magnetization of nickel in small particles.

Examples of particle diameters obtained in the two samples mentioned by the three methods, LHF, NR, and O_2 ads, respectively, were as follows: "Formate" sample:

6–7 nm; > 26 nm 66%, > 17 < 26 nm 26%,

> 11 < 17 0%, > 6 < 11 0%, < 5 nm 8%; 23 nm

"Impregnation" sample:

7–8 nm; > 26 nm 61%, > 17 < 26 nm

35%, > 11 < 17 nm 3%, > 6 < 11 1%, < 5 nm 0%; 31 nm

A study by Martin[16] has provided average particle diameters found for the metal in reduced nickel–silicas by the x-ray linewidth broadening, LLF, and NR methods. The results for a typical sample were, respectively, 6.3, 5.6, and 8.3 nm.

A more extensive study by Martin *et al.*[18–20] gives comparisons of results obtained on nickel–silica in several preparations, by low angle x-ray scattering, electron microscopy, x-ray line broadening, and both dc and ac variations of the NR method. The range of diameters observed was from about 2.0 to 30 nm. The range was, of course, somewhat different for the different experimental methods. In summary it may be said that the two variants of magnetic remanence measurements gave satisfactory agreement with each other, and with somewhat more detail for the ac method. Agreement was satisfactory between the magnetic methods and low angle scattering, with the former covering a lower diameter range. Agreement was also satisfactory between magnetic and linewidth broadening. For electron microscopy the agreement was less satisfactory although this method is probably the most searching. The discrepancies reported appear to be related to the greater sensitivity of the magnetic methods to very small particle sizes.

In conclusion it may be said that the magnetic methods, and especially the ac variant of the NR method, must be considered valuable, if not essential, aids in the granulometry of those systems of catalytic interest to which they are applicable. There have, for instance, been several applications of the methods described to nickel in various zeolites.[23] Tungler *et al.*[24] have, for instance, estimated M_s and T_C for nickel in Linde preparations 4A, 5A, 10X, and 13X, and observed the effect of adsorbed hydrogen. While these studies were not made at temperature below 298 K they nevertheless demonstrate the usefulness of the method. Tungler gives references to several other studies.

References

1. T. E. Whyte, Jr., *Catal. Rev.* **8**, 117 (1973).
2. H. Spindler, *Z. Chem.* **13**, 1 (1973).
3. B. Pulvermacher and E. Ruckenstein, *J. Catal.* **35**, 115 (1974).
4. J. L. Carter, J. A. Cusumano, and J. H. Sinfelt, *J. Phys. Chem.* **70**, 2257 (1966).
5. J. W. Cahn, *Trans. AIME* **209**, 1309 (1959).
6. R. E. Dietz and P. W. Selwood, *J. Chem. Phys.* **35**, 270 (1961).
7. W. Heukelom, J. J. Broeder, and L. L. van Reijen, *J. Chim. Phys.* **51**, 474 (1954).
8. W. Trzebiatowski and W. Romanowski, *Rocz. Chem.* **31**, 1123 (1957).
9. W. Trzebiatowski, "Catalysis and Chemical Kinetics," p. 193. Academic Press, New York, 1964.
10. J. T. Richardson, *J. Catal.* **21**, 122 (1971).
11. W. Romanowski, H. Dreyer, and D. Nehring, *Z. Anorg. Allg. Chem.* **310**, 286 (1961).
12. W. Romanowski, *Z. Anorg. Allg. Chem.* **351**, 180 (1967).
13. H. Dreyer, *Z. Anorg. Allg. Chem.* **362**, 233 (1968); **362**, 245 (1968).
14. W. H. E. Schwarz, *Z. Phys. Chem.*, (L), **247**, 265 (1971).
15. H. Charcosset, F. Figueras, L. de Mourgues, L. Tournayan, Y. Trambouze, and P. Weil, *J. Chim. Phys.* **65**, 1009 (1968).
16. G.-A. Martin, *J. Chim. Phys.* **66**, 140 (1969).

17. P. de Montgolfier and G.-A. Martin, *C.R. Acad. Sci. Ser. C* **273**, 1209 (1971).
18. G.-A. Martin, B. Moraweck, A.-J. Renouprez, G. Dalmai-Imelik, and B. Imelik, *J. Chim. Phys.* **69**, 532 (1972).
19. P. de Montgolfier, G.-A. Martin, and J.-A. Dalmon, *J. Phys. Chem. Solids* **34**, 801 (1973).
20. J. P. Candy, P. Fouilloux, G.-A. Martin, B. Blanc, and B. Imelik, *in* "Fine Particles," Second International Conference (W. E. Kuhn, ed.), pp. 218–223. Electrochem. Soc., Princeton, New Jersey, 1974.
21. L. Weil, *J. Chim. Phys.* **51**, 715 (1954).
22. P. O. Voznyuk and V. N. Dubinin, *Ukr. Fiz. Zh.* **19**, 160 (1974).
23. J. T. Richardson, *J. Catal.* **21**, 122 (1971).
24. A. Tungler, J. Petró, T. Mathé, G. Besenyei, and Z. Csűrös, *Acta Chim. (Budapest).* **82**, 183 (1974).

V

Magnetic Saturation Results for
H₂/Ni, H₂/Ni–Cu, H₂/Co,
and H₂/Fe

1. The Fractional Change of M_0

In this chapter there will be presented experimental results on the change of saturation magnetization when hydrogen is chemisorbed on each of the adsorbents named above. Later chapters will be devoted to discussion of these results and to those involving other adsorbates.

It has been shown, in previous chapters, that measurements of M must be made at low temperature and high field to yield reasonably accurate extrapolations to M_0. (This is especially true for adsorbents consisting of very small particles.) It will be repeated that, while these measurements are made at liquid-helium temperature, the adsorbate is, of necessity, admitted at much higher temperature. Virtually no vapor remains in the gas phase during actual measurements.

According to Eq. (1.9), in a sample of unit volume containing $n_p L$ particles of moment m_p, $M_0 = m_p n_p L$. Hence the fractional change of M_0 to M_0', namely, $\Delta M_0 / M_0$, caused by adsorbed molecules must be

$$\Delta M_0 / M_0 = \Delta(m_p n_p) L \tag{5.1}$$

We may define a quantity ϵ as the change in moment, of a particle, caused by the adsorption of one *atom* of hydrogen. It will be convenient to express the moments as Bohr magneton numbers $(\beta_p = m_p / m_B)$ in which case

$$\epsilon = (\Delta M_0 / M_0) n_p \beta_p / n(\text{H}) \tag{5.2}$$

and ϵ will be dimensionless. But in due course we shall expand the definition of ϵ to mean the change in moment of a particle of any adsorbent by an atom, or molecule, of any designated adsorbate. Equation (5.2) may, therefore, be written more precisely as

$$\epsilon_{Ni}(H) = (\Delta M_0/M_0) n(Ni) \beta(Ni)/n(H)$$

In recent work Martin and Imelik[1] have treated this important relation slightly differently, defining a number α as the change of Bohr magneton number due to the adsorption of one *molecule* of hydrogen. Hence $\alpha \simeq 2\epsilon$. We shall refer to another slight difference later.

2. Hydrogen on Nickel

Figure 20 shows saturation data obtained by Dietz and Selwood[2] on a sample of commercial nickel–kieselguhr. This had been reduced at 623 K for 12 hr, heated for several hours in helium at 873 K to cause a moderate increase of particle size, then cooled to the temperature of measurement which was 4.2 K. The weight of nickel in the sample was 0.1534 g. Data are also shown for the same sample after it had adsorbed 1.19 cm³ (STP) of hydrogen. The saturation magnetization of this sample prior to admission of the hydrogen was 97% of that of massive nickel. The mean particle radius derived from low-field data $(\bar{v}^2/\bar{v})$ was 6.4 nm.

From Fig. 18 the corrected galvanometer deflections (which vary directly with magnetization) show that $\Delta M_0/M_0 = -0.72/14.78 = -0.0487$.

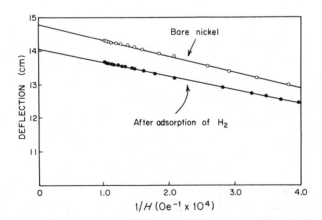

FIG. 20. The approach to M_s for a nickel–kieselguhr sample at 4.2 K before and after adsorption, at 298 K, of 7.76 cm³ (STP) H₂/g Ni.

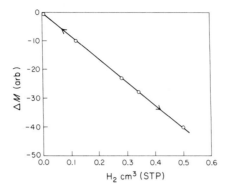

FIG. 21. The linearity of ΔM as a function of surface coverage of nickel by hydrogen (after Martin et al., Ref. 3).

We shall assume that $\beta(\text{Ni}) = 0.606$. Then ϵ may be calculated as follows:

$$\epsilon_{\text{Ni}}(\text{H}) = (\Delta M_0/M_0)\,n(\text{Ni})\beta(\text{Ni})/n(\text{H})$$

$$= \frac{-4.87 \times 10^{-2} \times 0.606 \times 0.1534 \text{ g} \times 22.4 \times 10^3 \text{ cm}^3\cdot\text{mol}^{-1}}{1.19 \text{ cm}^3 \times 2 \times 58.71 \text{ g}\cdot\text{mol}^{-1}}$$

$$= -0.72 \tag{5.3}$$

The average of several determinations on similar samples with moderately varying quantities of adsorbed hydrogen gave an average ϵ of -0.71. The average M_0 for these samples was 98.5% of that for massive nickel. The choice of moderately sintered samples for study was dictated by the convenient particle size which was small enough to adsorb a fairly large amount of hydrogen yet large enough to permit an accurate extrapolation to $1/H = 0$, to obtain M_0.

Extension of results, such as the above, to situations of changing surface coverage of the nickel by the hydrogen are not very easy to perform. The reason for this is that the sample must be heated to the temperature of adsorption between each of the measurements at liquid-helium temperature. Nevertheless, Martin et al.[3] have obtained such data, shown in Fig. 21, which is for hydrogen adsorbed on a basic nickel silicate (nickel antigorite) reduced, presumably, in the usual way. The mean nickel particle diameter determined magnetically was 6.0 nm. Adsorption took place at 300 K and magnetic measurements at 4.2 K. The nickel content was about 35 mg. The range of surface coverage extended, therefore, well below and above that reported by Dietz and Selwood.[2] It will be seen that ΔM is linear with the volume of hydrogen adsorbed. Martin et al.[3] report also that the adsorption and desorption isotherms are identical, and that our quantity

$\epsilon_{Ni}(H)$ = −0.73. The agreement for these systems is all that could be desired.

It is well known that some adsorbates change their mode of adsorption with time. There do not appear to have been any saturation studies of such possible effects on the hydrogen–nickel system. This refers, of course, to the time at, or near, adsorption temperature and not to the liquid-helium region. Results at low H/T to be described later make it unlikely that appreciable change occurs over a period of a few hours.

If adsorption is carried out at low temperature the hydrogen will be physically rather than chemically adsorbed. This certainly appears to be the case for adsorption much below 77 K. Martin and Imelik[1] have reported values of ϵ (given as $\alpha = 2\epsilon$) versus adsorption temperature over the range 173–723 K. These results shown in Fig. 22, were obtained for hydrogen adsorbed on a nickel–silica preparation obtained by the impregnation method and reduced at 893 K. It is obvious that for this system ϵ is independent of adsorption temperature over the indicated range.

The question of possible influences of particle size on $\beta(Ni)$ and on T_C have already been discussed (pp. 24–26). If any appreciable influences exist they would certainly be expected to have effects on ϵ. But such effects appear to be unlikely.

The question of contaminants, both in and on the adsorbent are potentially more serious. The possibility of dissolved impurities being present and being of such a nature as to affect ϵ seem to be readily dismissable. All workers in the field have found no foreign element present in the nickel in quantity sufficient to cause measurable change in the adsorptive properties, but the problem of surface cleanliness deserves careful consideration. Surface cleanliness is one of the most persistently debated questions in surface chemistry. On the one hand, it is pointed out that little quantitative information may be obtained relative to a surface the composition of which is unknown. On the other hand, it is obvious that solid surfaces used in

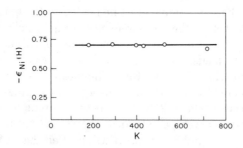

FIG. 22.   The independence of $\epsilon_{Ni}(H)$ as a function of chemisorption temperature (after Martin and Imelik, Ref. 1).

actual catalytic practice are far from being uncontaminated. There is some evidence that the metal surfaces used in the investigations under discussion are relatively free from surface contamination.

Dietz and Selwood[2] have shown that a pellet of nickel–silica evacuated to a minimum pressure of 10^{-4} N·m^{-2} may be no more contaminated than a nickel film at 10^{-8}. The reason for this is that the surface area of nickel metal in a typical nickel–silica pellet weighing, say, 1 g may be at least 10 m^2, while that of a typical film used in adsorption studies may be et least 10^3 times smaller. Furthermore, this difference in surface area may be accentuated by the nature of tt diffusion process necessary for a gas molecule to move inside the pellet. Gas molecules are able to reach the metal in the pellet only by passing the geometric boundary of the pellet, and this may be a surface of no more than 1 cm^2. The number of molecules entering the pellet per unit time may thus be less than the number striking the surface of a typical film. Yet once inside the pellet the vapor molecules find themselves faced with a metal surface many orders of magnitude larger than that of the film. To state this in another way, the metal–silica pellet acts as a "getter" for any gas molecule which has crossed its geometric boundary.

A rather different way to present the same argument is as follows: Let us say that the sample chamber has a dead space of 100 cm^3 and that this is held at a pressure of 10^{-4} N·m^{-2}. The total quantity of contaminating gas present is then far too small to make any measurable effect on a gram or two of nickel possessing a specific surface in the neighborhood of 10 m^2 or more. But the same quantity of gas could have a very serious effect on a film of which the surface was only 100 cm^2.

The above argument does not apply to contaminants which may not have been removed in the preparation stage or which may emerge from the silica support in the same manner as water may emerge from the glass on which a metal film is condensed. There are two lines of evidence which tend to show that neither of these sources of contamination is serious. The first is that monolayer coverage of nickel lowers the saturation magnetization by about 10% in a typical nickel–silica sample. The fact that $\beta(\mathrm{Ni})$ for the nickel in these samples is the same, within $\pm 1\%$, as that of pure, massive nickel suggests that no more than one-tenth of the surface could be contaminated.

There is quite another line of evidence which tends to confirm the views expressed above. It is well known that hydrogen is chemisorbed virtually instantaneously on nickel surfaces in the temperature range from about 123 K to several hundred degrees above room temperature. This is followed by a slower hydrogen sorption, the nature of which is still somewhat obscure but concerning which we shall have more to say later. Schuit and de Boer[4] have made the point that the slow effect, which is negligible on metal films, is also negligible on exhaustively reduced nickel–silica systems. There is no essential difference in the kinetics of adsorption on a film and on a supported metal, provided that the latter has been prepared with appropriate care. The inference is that if the film is free from contamination the supported metal is likewise free. This view receives confirmation from the experiment in which Schuit and de Boer deliberately contaminate the surface of the supported metal with oxygen, and show that this contamination causes a marked increase in the volume of "slow" hydrogen taken up, at the expense of the "instantaneous" hydrogen.

These several lines of evidence argue for the surface cleanliness of supported metal systems provided, of course, they have not been allowed to stand for any appreciable length of time after reduction, and especially if they have not been heated in vacuum or in inert atmosphere longer than is necessary to remove the residual adsorbed hydrogen after the reduction step. By contrast, the evidence that films, as ordinarily prepared, are similarly free from contamination is all circumstantial evidence. Films are sometimes handled in air after "protection" of the surface by evaporated silicon monoxide. The writer believes that this procedure is indefensible as actually consisting of gross and complete contamination of the surface. Silica gel, used as a metal support, always contains a trace of water. It might be thought that this water would emerge from the silica and contaminate the metal either as adsorbed water molecules or as oxide and hydrogen. This process probably occurs at elevated temperatures, but we shall later present some virtually conclusive evidence that progressive contamination does not readily occur at room temperature in a good vacuum system.

The possibility that the silica support may exert some influence on the electron distribution in small supported metal particles is one which cannot be dismissed lightly. This effect, if it exists, might be considered a kind of contamination by the support. Films should similarly show such an effect unless they are prepared, as rarely occurs, free from any glass or other surface on which the film is condensed. There is a fair amount of evidence dealing with this problem; we shall be concerned with possible influences of the support on ϵ. Interest in this area comes from the often-repeated observation that the catalytic activity of a supported metal depends, at least in part, on the nature of the support.

The first of these studies is that of Reinen and Selwood[5] who measured the saturation magnetizations up to 17 kOe and down to 2 K for nickel supported on silica gel and on high-area (so-called γ) alumina. The results obtained by Reinen are summarized in Table VI. The significantly low value of ϵ for the Ni–Al₂O₃ in the 2.6-nm sample could be due to some kind

TABLE VI

EFFECT OF SUPPORT ON ϵ_{Ni} (H)

Preparation	Ni (%)	v^2/v diam (nm)	$-\epsilon$ (av)
Ni–SiO₂ coppt	25	4.6	0.67
Ni–Al₂O₃ impreg	13.5	2.6	0.34
Ni–Al₂O₃ impreg sint	13.5	11.6	0.67
Ni–SiO₂ impreg	13.5	8.8	0.57

of electronic interaction between nickel and support, or it could be due simply to failure of the smaller particles to adsorb hydrogen, either from lack of complete reduction to metal or to protection from the hydrogen by the alumina.

The problem of possible support interaction on ϵ has also been studied by Martin et al.[3,6] who find that while on Ni–SiO₂ in several preparations ϵ is within the range given above, yet on a wide variety of supports including Al₂O₃, MgO, and various mixtures, the plots of ΔM_0 versus H₂ volume adsorbed are, for the most part, not linear and are not reversible on desorption. One example of many is shown in Fig. 23. This is for H₂ on a sample of Ni–α-Al₂O₃ containing 15.3 mg Ni reduced at 773 K and with an adsorption temperature 298 K. The data seem to fall into three approximately linear regions (one in adsorption and two in desorption), and at one stage ΔM is positive. The authors calculate ϵ for the three stages, and these have the following values: adsorption, -0.22; first desorption,

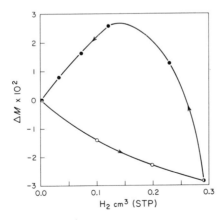

FIG. 23. A special case of an irreversible isotherm for H₂ adsorbed on Ni–Al₂O₃ (after Martin et al., Ref. 3).

−0.67; and second desorption, +0.6. The authors' explanation for these complex results involve the hypothesis of two different states of chemisorption. In view of this, further discussion will be postponed until we reach Chapter VII. It will be recalled (p. 52) that a positive change of magnetization has been observed on desorption of hydrogen from a nickel (metal)-containing zeolite. Further study including the measurement of ϵ on these systems would be useful.

3.   Hydrogen on Nickel–Copper Alloy

The only study of magnetization changes on this system appears to be that of Dalmon et al.[7] Silica-supported alloy particles were prepared by impregnation of high-area silica with mixed nitrate solutions to yield a final composition of 2 Ni + 1 Cu. Reduction was at 923 K and evacuation at 723 K followed by rapid cooling. Adsorption of H₂ was at 293 K and measurements at 4.2 and at 77 K. (These two temperatures gave almost identical results.) These measurements yielded $\beta(\text{Ni}) = 0.42$, which is consistent with previous work on this alloy, in the absence of any chemisorbed gas. The Curie temperature was also estimated to establish that no appreciable segregation of nickel on the surface may have occurred, although no Auger spectroscopy measurements for this purpose were performed.[8]

The results obtained are shown in Fig. 24. Adsorption and desorption points lie on the same straight line with an estimated $\epsilon_{\text{Ni}}(\text{H}) = -0.37$. Further work[9] by the same authors has extended these original results somewhat.

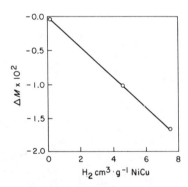

FIG. 24.   Magnetization-volume isotherm for H₂ on Ni-Cu alloy (after Dalmon *et al.*, Ref. 7).

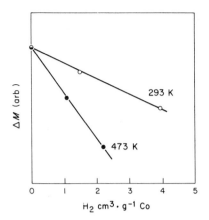

FIG. 25. Magnetization-volume isotherms for H_2 on Co (after Dalmon *et al.*, Ref. 7).

4. Hydrogen on Cobalt

Abeledo and Selwood[10] have reported on cobalt–silica prepared by impregnation and reduced at 523–673 K. The sample contained 10% Co. The particle diameter found from low-field data was 7.5 nm, and M_s was within 2% of the accepted value corresponding to $\beta(\text{Co}) = 1.7$. Hydrogen was adsorbed at 300 K and, on the basis of one experiment $\epsilon_{\text{Co}}(\text{H}) = -0.54 \pm 0.06$. The decrease of magnetization caused by hydrogen adsorbed over cobalt on a variety of supports has also been observed by Romanowski.[11]

In another study reported by Dalmon *et al.*[12] the sample was prepared by impregnation on silica, but with reduction at 773 K, and adsorption at 293 and 473 K. The magnetization results for three different volumes of hydrogen and at two temperatures are shown in Fig. 25. The diminution of M_s is linear. It is not stated if the results are reversible. From these data the authors find $\epsilon_{\text{Co}}(\text{H}) \simeq -0.17$ for adsorption at 293 K and -0.35 for adsorption at 473 K.

5. Hydrogen on Iron

Déportes *et al.*[13] have studied unsupported fine particles of iron prepared by cautious drying and reduction of hydrous iron(III) oxide. The highest temperature reached was 563 K at a pressure of $\sim 10^{-4}$ N·m^{-2}. The surface area $\simeq 10$ m^2·g^{-1}. Hydrogen was adsorbed at 373 K under $\sim 10^2$ N·m^{-2}.

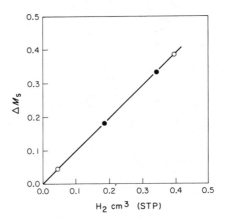

F$_{\mathrm{IG}}$. 26. Magnetization–volume isotherm for H$_2$ on Fe (after Déportes *et al.*, Ref. 13).

Magnetizations were measured at 4.2 K and up to 50 kOe with the use of a superconductive magnet. A sample completely reduced to metal adsorbed such a small quantity of hydrogen that extended measurements were not possible although an increase of ΔM_s was clearly indicated. A sample only 84% reduced took up about 0.45 cm³ H$_2$ per gram of iron. The data are shown in Fig. 26. The quantity $\epsilon_{\mathrm{Fe}}(H)$ was estimated to be $+ 1.85 \pm 0.2$. The saturation Bohr magneton number for iron free from adsorbed molecules is about 2.2.

The only other study on Fe/H$_2$ at conditions remotely approaching saturation appears to be that of Dumesic *et al.*[14] who used a maximum field of about 3.7 kOe and a minimum temperature of 77 K. The iron was supported on magnesia with metal particles in the 1.5-nm range. At moderately elevated temperatures the effect of adsorbed hydrogen on the magnetization was no more than a 2% increase. This gives $\epsilon_{\mathrm{Fe}}(H) \leq 0.1$. A larger positive change, attributed by the authors to a diminishing anisotropy barrier, was found at lower temperatures.

In all this work difficulties arise from the poor reducibility of the iron to metal. One further study on Fe/H$_2$, at low field, will be described in Chapter VII. Gittleman *et al.*[15] have recently reinvestigated some of the magnetic properties of nickel–silica systems and have used co-sputtering of Ni and SiO$_2$ (and Al$_2$O$_3$) to prepare the samples. If this preparation method can be extended to Co–SiO$_2$ and especially to Fe–SiO$_2$ it would solve some of the problems hitherto encountered with these systems. An Fe–Ni–SiO$_2$ co-sputter film has actually been obtained,[16] but it is not clear to what degree the metal surfaces in these preparations may be accessible to vapor molecules.

References

1. G.-A. Martin and B. Imelik, *Surface Sci.* **42**, 157 (1974).
2. R. E. Dietz and P. W. Selwood, *J. Chem. Phys.* **35**, 270 (1961).
3. G.-A. Martin, G. Dalmai-Imelik, and B. Imelik, *in* "Proceedings of the Second International Conference on Absorption–Desorption Phenomenon, Florence, 1971" (F. Ricca, ed.), p. 433. Academic Press, New York, 1972.
4. G. C. A. Schuit and N. H. de Boer, *Rev. Trav. Chim.* **70**, 1080 (1951).
5. D. Reinen and P. W. Selwood, *J. Catal.* **2**, 109 (1963).
6. G.-A. Martin, N. Ceaphalan, P. de Montgolfier, and B. Imelik, *J. Chim. Phys.*, (10), 1422 (1973).
7. J.-A. Dalmon, G.-A. Martin, and B. Imelik, *Surface Sci.* **41**, 587 (1974).
8. C. R. Helms, *J. Catal.* **36**, 114 (1975).
9. J.-A. Dalmon, G.-A. Martin, and B. Imelik, *Jap. J. Appl. Phys.*, Suppl. 2, Pt. 2. (1974).
10. C. R. Abeledo and P. W. Selwood, *J. Chem. Phys.* **37**, 2709 (1962).
11. W. Romanowski, *Chem. Stosow.* **2**, 225 (1961).
12. J.-A. Dalmon, G.-A. Martin, and B. Imelik, "Thermochimie," Colloq. No. 201, pp. 593–600. Centre Nationale de la Recherche Scientifique, Marseille, 1971.
13. J. Déportes, J.-P. Rebouillat, R. Dutartre, J.-A. Dalmon, and G.-A. Martin, *C.R. Acad. Sci. Ser. C* **276**, 1393 (1973).
14. J. A. Dumesic, H. Topsøe, J. H. Anderson, and M. Boudart, *Surface Sci.* (to be published).
15. J. I. Gittleman, B. Abeles, and S. Borowski, *Phys. Rev. B* **9**, 3891 (1974).
16. J. J. Hanak and J. I. Gittleman, *AIP Conf. Proc. Magnetism and Magnetic Materials* **10**, 961 (1972).

VI

Magnetization Measurements
at Low M/M_0

1. The Permeameter

However satisfying the direct measurement of saturation magnetization may be, it cannot be overlooked that few catalytic reactions proceed with measurable velocity at 4.2 K. Surface chemistry in this area will best be served by physical measurements under conditions that favor chemical reactivity. This means that measurements should generally be made at room temperature and higher, and at pressures up to and above 1 atm. If the large magnet may be dispensed with, so much the better. Fortunately, these conditions may be met. If the results cannot always be expressed in unequivocal terms they are nevertheless attended with flexibility, and they yield a wealth of information. On the other hand, there are severe restrictions on measurements at relatively low H/T, corresponding to $M \ll M_0$. If these restrictions are ignored the results will be meaningless.

Magnetizations on samples exhibiting superparamagnetism, under the conditions of measurement, were first made by Heukelom et al.[1] These measurements were made by an induction method, or permeameter. This is essentially a step-down transformer with the sample placed in a secondary coil that, in turn, is placed in a primary. When the current in the primary is turned on or off, or reversed, a charge is induced in the secondary and is measured on a ballistic galvanometer, or other current integrating device. The induced charge is proportional to the magnetization of the sample. The field is rarely over a few hundred oersteds. The temperature of the sample and the pressure of the gas over it may be adjusted at will. But for meaningful measurements the sample must be superparamagnetic under the conditions of the experiment.

It is possible to use alternating current in the primary solenoid.[2] This adds substantially to convenience and to the possibility of observing very rapid changes in M. But ac also adds further restrictions to the method. The following description is with reference to the ac permeameter, but with simple changes it may be adapted to dc.

It is convenient to have the secondary made up in two coils arranged coaxially, and several centimeters apart. The sample is placed in the core of one coil and the other, which is connected in opposition to the first, is left empty, or it may be filled with sample but sealed off from adsorbable gas. This arrangement makes possible a considerable increase in sensitivity.

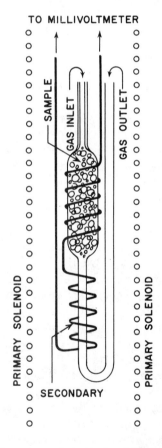

FIG. 27. Secondary coils and sample in the low-frequency ac permeameter.

FIG. 28. The secondary coil assembly for the permeameter.

The secondary assembly is shown in Fig. 27. Each coil of the secondary has about 40 turns. It is quite convenient to construct the secondary so that it may be heated to the reduction temperature, namely, about 673 K. This obviates the difficulty of resetting the sample during the course of any series of measurements.

The wire used for the secondary should be nonmagnetic and not readily oxidized at moderately elevated temperature. Chromel A is satisfactory for this purpose. This is wound on a threaded form which is then fired to give a strong ceramic-like body. The secondary assembly is mounted on aluminum rods which serve both as supports and as electrical conductors, leading to a vacuum-tube millivoltmeter. It will be noted that the secondary has quite a low impedance. This is intentionally so to make the millivoltmeter relatively stable to electrical transients. The complete secondary is shown in Fig. 28. The sample chamber is sealed into the adsorption train, Fig. 29, prior to reduction and evacuation. About 5 g of pelleted sample is convenient.

The secondary coils and sample are surrounded by a primary coil which may be raised or lowered as desired. The primary consists of about 5000 turns of insulated No. 15 copper wire wound in about 20 layers, some 30 cm long, on a brass core of about 7.5 cm internal diameter. The large core is for introduction of temperature control equipment around the sample. The primary coil is operated at about 0.75 A, 60 Hz alternating current. The power supply is stabilized by a voltage regulator. If long runs are planned, it is advantageous to cool the primary core by a few turns of copper tubing carrying running water.

The choice of stabilized 60 Hz requires some explanation.[3] It is essential that the sample particles should have a relaxation time τ short enough so that M will reach a maximum in a time short compared with the applied power cycle. A typical nickel–silica sample fulfills this requirement at room temperature, but not by a very wide margin.

A slightly modified design of ac permeameter is based on the original design of Geus et al.[4] described by Artyukh et al.[5] Some details are shown in Fig. 30.

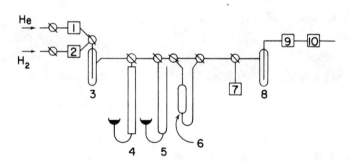

FIG. 29. Gas handling system for low-frequency ac permeameter studies: (1) helium purification, (2) hydrogen purification, (3) silica gel trap, (4) gas buret, (5) mercury manometer, (6) sample, (7) McLeod gauge, (8) trap, (9) diffusion pump, (10) mechanical pump.

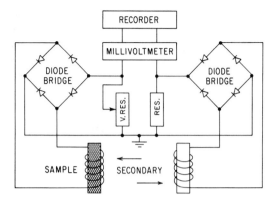

FIG. 30. Alternative permeameter design (after Artyukh *et al.*, Ref. 5).

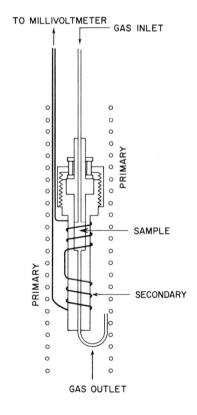

FIG. 31. Permeameter secondary and sample holder for elevated pressures.

For measurements at elevated pressures a modified permeameter has been designed by Vaska and Selwood.[6] The sample holder consists of a stainless-steel nonmagnetic bomb around the outside of which is wound the secondary coil. This design, shown in Fig. 31 which results in some loss of sensitivity, permits the use of pressures up to about 100 atm.

2. General Considerations

It is readily shown, as might be expected from the geometry of the apparatus, that the emf produced in the secondary assembly is directly proportional to the magnetization M of the sample.* But this is true only, as pointed out above, if the sample exhibits true superparamagnetism under the experimental conditions. Dietz[3] shows that two tests may conveniently be made to establish this condition. The first is that as M should vary inversely as the absolute temperature and directly as the square of the spontaneous magnetization, we may write

$$\frac{E(T_1) - E_0}{E(T_2) - E_0} = \frac{T_2 M_{sp}^2(T_1)}{T_1 M_{sp}^2(T_2)} \qquad (6.1)$$

where $E(T_1)$, $E(T_2)$ are the secondary emf readings obtained for a sample at temperatures T_1 and T_2, E_0 is the null emf before insertion (or reduction) of the sample, and $M_{sp}(T_1)$ and $M_{sp}(T_2)$ are the spontaneous magnetizations at T_1 and T_2, respectively.

Similarly, the emf should be directly proportional to the primary current:

$$\frac{E(i_1) - E_0(i_1)}{E(i_2) - E_0(i_2)} = \frac{i_1}{i_2} \qquad (6.2)$$

where i is the current through the primary.

A third possible test relates to the slope of the magnetization–volume isotherm and will be discussed in the following chapter.

If it is so desired the secondary emf may be recorded directly with a 1-mV recorder. This is connected through appropriate rectification and an isolating transformer.

Heating of the sample is done with a tube furnace which slips through the core of the primary. For reductions the primary is moved out of position, but the furnace may also be used as a thermostat for measurements above room temperature. The furnace is heated with direct current under these

---

* The null emf, prior to insertion of the sample, may be made negligible or nearly so, by adjusting the relative positions of primary and secondary.

conditions. Owing to the relation of magnetization to temperature it is necessary to thermostat the sample to about ±0.1 K while data for an isotherm are being obtained. This is done by flowing water from a bath at constant temperature through a cylindrical container surrounding the sample. For measurements below room temperature a Dewar flask fits into the core of the primary.

This section will be concluded with a brief description of general procedure which may be followed for obtaining magnetization–volume isotherms for, say, hydrogen on nickel–silica. The sample in the form of $\frac{1}{8}$-inch pellets is sealed into the sample chamber in such a manner as to fill, as completely as convenient, the core of one of the two opposing secondary solenoids. Five grams of a typical sample containing 40% nickel, as metal, will give rather more than enough precision.

The assembly for normal gas handling is shown in Fig. 29 and for elevated pressures in Fig. 32.

The sample is reduced in flowing, purified hydrogen at elevated temperature for about 12 hr, evacuated down to a pressure of about 10^{-4} N·m^{-2} at 673 K for 2 hr, and then allowed to cool to the temperature of measurement. If this temperature is much below 273 K, it may be necessary to introduce a trace of helium to hasten the attainment of thermal equilibrium.

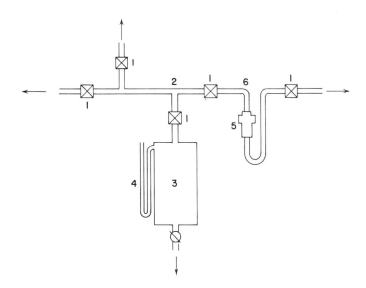

Fig. 32. Gas buret system for obtaining magnetization-volume isotherms at elevated pressures. (1) steel valves, (2) steel pressure tubing, (3) glass exhaust chamber serving as buret, (4) open-arm manometer, (5) adsorbent chamber, (6) stainless-steel tubing $\frac{1}{16}$ in. id (after Naska and Selwood, Ref. 6).

After the sample has reached the desired temperature the secondary emf is obtained for a fixed primary current. Increments of hydrogen are then admitted to the sample. After each increment it is essential to wait until the sample returns to temperature. At low surface coverage, when the heat of adsorption is still quite high, it may be necessary to wait 10 min or more until the heat liberated at each addition of hydrogen is dissipated.

Through successive quantitative additions of hydrogen, after each of which the pressure and the secondary emf are found, there will be obtained data from which may be plotted the magnetization–volume isotherm. The fractional change of magnetization is given by

$$\frac{\Delta M}{M} = \frac{\Delta(E - E_0)}{E - E_0} \tag{6.3}$$

In view of this no further calibration is necessary.

Pressure–volume isotherms are, of course, obtained simultaneously. It is convenient to plot the final result as the fractional change of magnetization against cm^3 (STP) of hydrogen adsorbed per gram of metal. A virtue of this experimental method is that the correction for gas in the dead space may, by proper design, be kept quite moderate. Except at elevated pressures the volume of gas in the dead space may rarely exceed 20% of the total volume of adsorbate admitted to the sample chamber.

3.   Theory of Low M/M_0 Measurements

If all the particles of a sample exhibiting superparamagnetism have the same size, as never occurs in practice, then under conditions such that $M/M_s \ll 1$, the magnetization is essentially that of a paramagnetic, Eq. (1.2), and may be written

$$M = n_p L (M_{sp} v)^2 H / 3kT \tag{6.4}$$

($n_p L$ being the number of particles of volume v in the sample).

It may be wondered why we concern ourselves with an ideal situation that no one has yet achieved, namely, a specimen containing particles of one size only. The reason is that it is easier to understand the derivation for the isotherm if this ideal situation is treated first. To a degree real catalyst samples behave in part as if they actually contain such particles. But there is another implied assumption in what follows. This is that adsorbed molecules are distributed uniformly over the surface regardless of any differences in adsorbent particle size. We shall examine this assumption in detail later.

If hydrogen is adsorbed on a particle of moment $M_{sp}v$, the moment may be expected to change to become $M_{sp}v - n(\text{H})L\epsilon m_\text{B}$, where $n(\text{H})$ here is the number of moles of hydrogen atoms adsorbed on the *particle*. The magnetization M of the sample thus altered by the presence of adsorbed atoms will be

$$M' = \frac{n_\text{p}L[M_{sp}v - n(\text{H})L\epsilon m_\text{B}]^2 H}{3kT} \tag{6.5}$$

and the fractional change of magnetization will be

$$\frac{\Delta M}{M} = \frac{[n_\text{p}(M_{sp}v - n(\text{H})L\epsilon m_\text{B})^2 H/3kT] - [n_\text{p}(M_{sp}v)^2 H/3kT]}{n_\text{p}(M_{sp}v)^2 H/3kT}$$

$$= \frac{-2n(\text{H})L\epsilon m_\text{B}}{M_{sp}v} + \frac{[n(\text{H})L\epsilon m_\text{B}]^2}{M_{sp}v} \tag{6.6}$$

From Eq. (6.6) we see that the initial fractional change of magnetization caused by adsorbed molecules varies directly as the number of such molecules taken up by a particle, but inversely as the volume of the individual particles. While the temperature does not appear directly in the above expression, it will be recalled that M_{sp} diminishes somewhat with increasing temperature. One consequence of this is that the magnetization–volume isotherm should have a somewhat steeper slope as the temperature is raised.

It will be noted also that the second, squared, term will cause the isotherm to bend away from the volume axis as surface coverage is increased. In practice, $\Delta M/M$ rarely exceeds 30%. A consequence of this is that the isotherm should be approximately a straight line, the slope of which is independent of particle volume. A given mass of smaller particles will, of course, adsorb more hydrogen than will the same mass of larger particles. The magnetization–volume isotherm for a more finely dispersed system will be longer than that for a coarser system but it should have the same slope. A typical theoretical isotherm for a monodispersion is shown in Fig. 33.

Next we shall present the treatment developed by Dietz and Selwood[7] for samples containing a distribution of particle diameters. From Eq. (6.5):

$$M' = \frac{H}{3kT} \sum_\text{p} n_\text{p}L(m_\text{p} - \Delta m_\text{p})^2 \tag{6.7}$$

n_p being the moles of particles of radius r and $\Delta m_\text{p} = m_\text{p}' - m_\text{p}$, namely, the change of particle moment caused by an adsorbate.

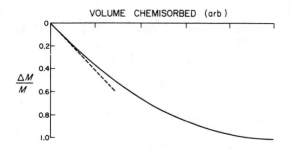

FIG. 33.   Theoretical dependence of $\Delta M/M$ on volume of adsorbate, for a mono-dispersed adsorbent.

Then the fractional change in magnetization is

$$\frac{\Delta M}{M} = \frac{\sum n_\text{p} L (\Delta m_\text{p})^2 - 2 \sum n_\text{p} L \,\Delta m_\text{p}}{2 n_\text{p} L m_\text{p}} \qquad (6.8)$$

If the particles under consideration are spherical

$$\Delta m_\text{p} = 4\pi r^2 n_\text{s}(\text{H}) L \theta \epsilon m_\text{B} \qquad (6.9)$$

where $n_\text{s}(\text{H}) L$ is the number of hydrogen atoms adsorbed per unit surface and θ the fractional surface coverage.

It will be noted that

$$\frac{n_\text{s}(\text{H}) L \theta A_\text{p}}{M_\text{sp} v} = \frac{\Delta M_\text{s}}{M_\text{s}} \qquad (6.10)$$

where A_p is the total surface area of the particle, whence, by substitution in Eq. (6.9),

$$\Delta m_\text{p} = \frac{\Delta M_\text{s}}{M_\text{s}} \cdot \frac{(M_\text{sp} V)}{A_\text{p}} \, 4\pi r^2$$

$$= \frac{\Delta M_\text{s}}{M_\text{s}} \cdot \frac{(M_\text{sp} V)}{A_\text{p}} \, 4\pi \left(\frac{3v}{4\pi}\right)^{2/3} \qquad (6.11)$$

We transform n_i into a continuous volume distribution function:

$$n_i = \frac{f(v) \, dv}{v} \qquad (6.12)$$

where

$$\int_0^\infty f(v) \, dv = V \qquad (6.13)$$

Furthermore, expressing A_p in terms of the continuous distribution function

$$A_p = \int_0^\infty 4\pi r^2 \frac{f(v)\ dv}{v}$$

$$= 4\pi (\tfrac{3}{4}\pi)^{2/3} \int_0^\infty \frac{f(v)\ dv}{v^{1/3}} \tag{6.14}$$

and substituting into Eq. (6.8), we have

$$\frac{\Delta M}{M} = -\alpha\left(\frac{\Delta M_s}{M_s}\right) + \delta\left(\frac{\Delta M}{M}\right)^2 \tag{6.15}$$

where

$$\alpha = 2\ \frac{\displaystyle\int_0^\infty v^{2/3}f(v)\ dv \int_0^\infty f(v)\ dv}{\displaystyle\int_0^\infty v^{-1/3}f(v)\ dv \int_0^\infty fv(v)\ dv} \tag{6.16}$$

and

$$\frac{\alpha}{\delta} = 2\ \frac{\displaystyle\int_0^\infty v^{2/3}f(v)\ dv \int_0^\infty v^{-1/3}f(v)\ dv}{\displaystyle\int_0^\infty f(v)\ dv \int_0^\infty v^{1/3}f(v)\ dv} \tag{6.17}$$

The symmetry of the ratio of integrals suggests that both α/δ and α are both approximately equal to 2. Experimental results obtained in this writer's laboratory confirm this view.

Table VII shows values of α and α/δ calculated for certain arbitrary particle size distributions.[3] (Additional values of α are given by Reinen and Selwood.[8]) It is, therefore, clear that provided the sample exhibits true superparamagnetism the low-field method gives a change $\Delta M/M$, which is proportional to the change of saturation magnetization. We note also that the second, squared, term may be neglected if $\Delta M/M$ is not large. Equation (6.15) reduces to Eq. (6.6) for a sample in which all adsorbent particles are the same size. This is true because, as $n(H)L = n_s(H)L\theta A_p$, it follows that

$$\Delta M_s/M_s = -n(H)L\epsilon m_B/M_{sp}v.$$

TABLE VII

THE QUANTITIES α AND α/δ, EQ. (6.15), CALCULATED FOR CERTAIN
ARBITRARY DISTRIBUTIONS OF PARTICLE SIZE[a]

Distribution	α	α/δ
Delta function	2.000	2.000
Infinitely wide rectangle	1.600	2.400
Maxwellian (volumes)	1.867[b]	2.100
Maxwellian (radii)	1.471	2.500

[a] Reference 3.   [b] Incorrectly given in Ref. 3.

The case of a sample in which there are two sizes of particles and on which, at equilibrium fractional coverage of the surfaces are the same, is easily considered. From Eq. (6.16) we may write, for the two assemblies

$$\frac{\Delta M_1/M_1}{\Delta M_2/M_2} = \frac{-2n(\mathrm{H})_1 L \epsilon m_{\mathrm{B}}/M_{\mathrm{sp}} v_1}{2n(\mathrm{H})_2 L \epsilon m_{\mathrm{B}}/M_{\mathrm{sp}} v_2} \tag{6.18}$$

and from Eq. (6.4)

$$\frac{M_1}{M_2} = \frac{n_{\mathrm{p}1} v_1^2}{n_{\mathrm{p}2} v_2^2} \tag{6.19}$$

so that

$$\frac{\Delta M_1}{\Delta M_2} = \frac{n(\mathrm{H})_1 n_{\mathrm{p}1} v_1}{n(\mathrm{H})_2 n_{\mathrm{p}2} v_2} \tag{6.20}$$

$n(\mathrm{H})_1$ and $n(\mathrm{H})_2$ being the moles of H atoms adsorbed on particles of volumes v_1 and v_2, respectively.

The final problem to be dealt with in this section involves the case of nonuniform surface coverage. This may occur from inaccessibility of some adsorbent particles, or of accessibility so far reduced as to require very long times for equilibrium to be attained. There is also the probability that some particles, because of highly irregular surface, may adsorb certain molecules by a mechanism different from that on more nearly smooth surfaces. It must be pointed out that these are basic difficulties encountered in every study of practical catalyst surfaces. The problems are not unique to the magnetic method, but they lie at the heart of any real understanding of catalytic action on solid surfaces.

This problem has been considered by Martin et al.[9] To avoid confusion that might arise from the different symbols used by these authors we shall merely present the final conclusion [neglecting the squared term in Eq.

(6.15)]. This conclusion is that instead of $\Delta M/M = \alpha(\Delta M_s/M_s)$ there should be another term such that

$$\Delta M/M = - (\Delta M_s/M_s)(1 + \phi) \qquad (6.21)$$

where ϕ is a function of the particle diameter distribution and of the preferential adsorption on some particles. (Note that α is used by Martin *et al.* in a different meaning from that used in this book.) Equation (6.18) shows that the magnetization of a sample varies linearly with the quantity of gas adsorbed, but the relationship at high M/M_0 involves simply the number ϵ while that at low M/M_0 involves $\epsilon(1 + \phi)$. A comparison of data obtained at high and at low M/M_0 permits, therefore, an estimation of ϕ and, if the particle size distribution is known, it permits an estimation of surface coverage variations with changing particle size. The implications of these possibilities will be described in the following chapters.

References

1. W. Heukelom, J. J. Broeder, and L. L. van Reijen, *J. Chim. Phys.* **51,** 474 (1954).
2. P. W. Selwood, *J. Amer. Chem. Soc.* **78,** 3893 (1956).
3. R. E. Dietz, Doctoral Dissertation, p. 111. Northwestern University, Evanston, Illinois, 1960.
4. J. W. Geus, A. P. P. Nobel, and P. Zwietering, *J. Catal.* **1,** 8 (1962).
5. E. N. Artyukh, N. K. Lunev, and M. T. Rusov, *Kinet. Catal.* **13,** 663 (1972) [Engl. transl.].
6. L. Vaska and P. W. Selwood, *J. Amer. Chem. Soc.* **80,** 1331 (1958).
7. R. E. Dietz and P. W. Selwood, *J. Chem. Phys.* **35,** 270 (1961).
8. D. Reinen and P. W. Selwood, *J. Catal.* **2,** 109 (1963).
9. J.-A. Martin, Ph. de Montgolfier, and B. Imelik, *Surface Sci.* **36,** 679 (1973).

VII

Low–Field Results for H₂/Ni, H₂/Co, and H₂/Fe

1. Magnetization–Volume Isotherms for H_2/Ni

Figure 34 shows the fractional change of magnetization, $\Delta M/M$, plotted against volume of hydrogen adsorbed per gram of nickel present. The pressure–volume isotherm obtained simultaneously is also shown. These data were obtained on a nickel–kieselguhr sample containing about 50% nickel, reduced in hydrogen at 673 K for 12 hr, then evacuated to about 10^{-4} N·m⁻² for 2 hr, and allowed to cool in vacuum to room temperature. Magnetic measurements, with progressive increments of hydrogen, were made at room temperature on the ac permeameter. Data of this kind may be obtained for a wide variety of preparations,[1–4] but always with the requirement that the sample must pass the tests for superparamagnetism under the experimental conditions.

If, after the data for plotting a magnetization–volume isotherm have been obtained, the sample is evacuated, it will be found that at room temperature approximately one-third of the hydrogen may be desorbed. Prolonged evacuation removes a negligible additional volume of adsorbate. Within this limited region the magnetization is recovered linearly with the hydrogen; the isotherm is, therefore, reversible throughout the region in which hydrogen may be pumped off at room temperature. Studies[2] have shown that all the hydrogen originally admitted may, within experimental error, be recovered by evacuation for about 2 hr at 633 K. If, after this treatment, the sample is cooled to room temperature, it will generally be

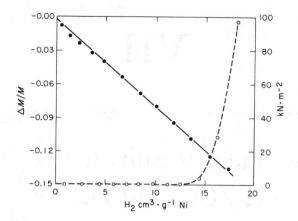

FIG. 34. Magnetization-volume ($\bullet$) and corresponding pressure-volume ($\bigcirc$) isotherms obtained at room temperature for hydrogen adsorbed on nickel-kieselguhr.

found that the magnetization is slightly greater than it was at the same temperature before admission of the hydrogen. The reason for this increase, which occurs progressively with every adsorption–desorption cycle, seems to be that the nickel particles slowly increase in size. But whether this is due to the heating necessary to desorb the hydrogen, or to some irreversible change related to the presence of the hydrogen, is not known. This effect concerns us only when it becomes necessary to compare the isotherm slope for hydrogen with that of some other adsorbate.

We shall consider whether the slope of the isotherm agrees with the prediction of Eq. (6.7) with due regard for the fact that real samples exhibit a considerable range of particle size. As $\Delta M/M$ does not exceed 15% the second, squared, term in Eq. (6.7) may be neglected. It is therefore obvious that the isotherm is, as predicted, a straight line. Then

$$\frac{\Delta M}{M} = \frac{-2n(\mathrm{H})\, L\epsilon m_{\mathrm{B}}}{M_{\mathrm{sp}} v} \tag{7.1}$$

Multiplying numerator and denominator by the number of adsorbent particles present in the sample, that is, by $n_{\mathrm{p}} L$:

$$\frac{\Delta M}{M} = \frac{-2n(\mathrm{H})\, L\epsilon m_{\mathrm{B}} n_{\mathrm{p}} L}{M_{\mathrm{sp}} v n_{\mathrm{p}} L}$$

$$= \frac{-2n(\mathrm{H})\, L\epsilon m_{\mathrm{B}}}{M_{\mathrm{sp}} V} \tag{7.2}$$

where $n(\mathrm{H})\, L$ is the number of hydrogen atoms adsorbed on the sample and V is the volume of ferromagnetic adsorbent. Then for a typical nickel-silica adsorbing say 10 cm³ (STP) H_2/g Ni at 298 K, we have

$$\frac{\Delta M}{M} = \frac{-2 \times 2\,\mathrm{mol(H)} \cdot \mathrm{mol}^{-1}(\mathrm{H_2}) \times 6.02 \times 10^{23}\,\mathrm{mol}^{-1}}{22.4 \times 10^3\,\mathrm{cm}^3 \cdot \mathrm{mol}^{-1}}$$

$$\times \frac{10\,\mathrm{cm}^3 \times 0.72 \times 0.927 \times 10^{-2}\,\mathrm{Oe} \cdot \mathrm{cm}^3 \times 8.9\,\mathrm{g} \cdot \mathrm{cm}^3}{4.85 \times 10^2\,\mathrm{Oe} \times 1\,\mathrm{g}}$$

$$= -0.13 \text{ (calc)}$$

From Fig. 34 it will be seen that, for the particular sample and conditions indicated, $\Delta M/M$ is about -0.08. Various authors have reported values for α [Eq. (6.15)]. For instance, Geus et al.[5] give a value of 2.02. Variations from sample to sample are to be expected because, as shown in Table VII, $\alpha = 2.0$ for uniform particle size only. Few, if any, samples fall into this class.

2. Temperature of Adsorption and Temperature of Measurement

Magnetization–volume isotherms are dependent on both the temperature of adsorption and on the temperature at which the magnetic measurements are made.

Equation (7.2) shows that the slope of the isotherm should vary inversely as the spontaneous magnetization. Inasmuch as the spontaneous magnetization is a function of temperature, we may expect the isotherm slope to increase moderately with increasing temperature, even in the room-temperature region. Reference to Fig. 2 will show that M_{sp} decreases slowly, then more rapidly, as the temperature is raised. (This change should not be confused with the fact that the magnetization M varies inversely as the absolute temperature for any sample exhibiting superparamagnetism.)

Figure 35 shows isotherms[6] for hydrogen on nickel–silica at 195 and 479 K. These isotherms were obtained under conditions as nearly identical as possible except for the temperature change. The hydrogen was adsorbed at room temperature. The ratio of isotherm slopes is 0.71, increasing with increasing temperature as predicted. The inverse ratio of spontaneous magnetizations at these two temperatures (from Fig. 2) is 0.69. It will, however, be recalled that very small particles of nickel have a Curie temperature a little lower than that (631 K) of pure massive nickel. For particles of the size under consideration, the difference is about 50 K, and this results in a corrected magnetization curve for which M_{sp} (479 K)$/M_{\mathrm{sp}}$ (195 K) $= 0.73$, which is a very satisfactory agreement. The most significant feature of this agreement is, however, that it establishes the invariance of ϵ over a wide temperature range. Additional data on the

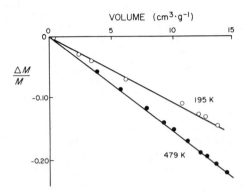

FIG. 35. Magnetization–volume isotherms at two different temperatures for hydrogen on nickel–silica.

effect of measurement temperature have been reported by Geus *et al.*[5] Again, if M_{sp} is corrected for the temperature change there is no significant change in the isotherm slope. [The same paper clearly shows the effect of the second, squared, term, Eq. (6.6), on the form of the isotherm at higher surface coverage.]

It is well known that the mode of adsorption of hydrogen on nickel is, to a degree, dependent on the temperature of adsorption.[7] At quite low temperatures the mechanism is primarily physical, while above about 123 K, this goes over to chemisorption. Some confirmatory evidence[3] concerning this view is obtained by the admission of hydrogen to a reduced nickel–silica sample at 77 K, as shown in Fig. 36. The volume of hydrogen adsorbed is very large, but the change of magnetization is negligible after a very small amount of hydrogen has been taken up. This result is in agreement with the observations of Sadek and Taylor,[8] who showed that, on a catalyst and conditions similar to those used in this investigation, of the 8.17 cm³ of H_2 adsorbed (with rising temperature) per gram Ni, 7.70 cm³ was physically adsorbed and only 0.47 cm³ chemisorbed. With falling temperature the respective volumes were approximately equal. This view receives further confirmation from the observation by Schuit *et al.*[9] that very little hydrogen–deuterium exchange occurs on nickel–silica at 77 K.

It is to be noted[3] that if the sample that holds a rather large volume of van der Waals hydrogen at 77 K is allowed to warm to room temperature, there occurs an abrupt loss of magnetization as the transition occurs from van der Waals to chemisorbed hydrogen. These results are, of course, in excellent agreement with those previously reached by various workers from several points of view. The abruptness of the change is caused by the self-accelerating effect of the heat of chemisorption.

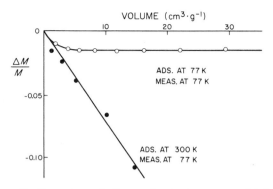

Fig. 36. Magnetization-volume isotherms for hydrogen on nickel–silica, showing dependance of adsorption mechanism on temperature of adsorption.

3. Elevated Pressure and Preadsorbed Molecules

It is sometimes stated that the surface coverage of nickel by chemisorbed hydrogen is virtually complete at a fraction of a millimeter pressure. This question is susceptible to examination by the magnetic method.

Figure 37 shows a magnetization–volume isotherm, obtained by Vaska and Selwood,[10] for hydrogen on nickel–silica carried up to about 70

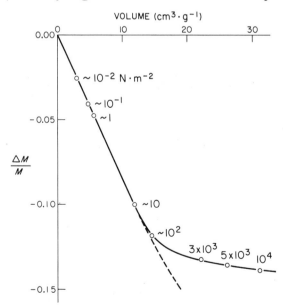

Fig. 37. Magnetization–volume isotherm at room temperature for hydrogen on nickel–silica up to about 70 atm. The dotted line represents the high-pressure part of the isotherm after appropriate corrections.

atm. At the same pressure the effect of helium is virtually negligible. When the results shown in Fig. 37 were first obtained, the effect of the squared term in Eq. (6.6) was not appreciated. One should, therefore, attribute a moderate part of the turning away from the volume axis to the squared term. Some part may be due to the "slow" sorption of hydrogen which, as is well known, always occurs in such systems, and which tends to become less slow as the pressure is raised. But these various considerations do not alter the primary conclusion which is that surface coverage is not complete on this system at atmospheric pressure. Coverage is essentially complete at 100 atm—in a typical run, the volume of chemisorbed hydrogen (as determined magnetically) was 17.6 cm³ per gram Ni at 23.4 atm and 17.9 cm³ at 71 atm.

The other point to be mentioned in this section is the possible effect of preadsorbed molecules on the magnetization–volume isotherm for hydrogen. In later chapters we shall make use of the technique of blocking certain fractions of the metal surface area by preadsorption with a different molecule. If the preadsorbed molecule is reactive, such as ethylene, then the hydrogen uptake at appropriate temperature, and the slope of the magnetization–volume isotherm, become complicated by the chemical changes that may occur on the surface. At 195 K preadsorbed benzene does not react with hydrogen. For this experiment the bare surface is exposed to benzene vapor at or slightly below room temperature. The sample is then cooled to 195 K and the hydrogen is admitted. The result is that while the volume of hydrogen adsorbed is decreased in proportion to the quantity of benzene preadsorbed, the slope of the isotherm is not affected by the presence of the benzene. (H₂ is chemisorbed on a bare Ni surface at 195 K.) This result, which is in general true of all preadsorbates (provided

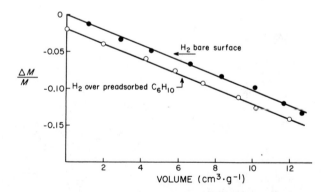

FIG. 38. Magnetization–volume isotherms for hydrogen on nickel–silica at 195 K before and after the preadsorption of cyclohexene.

that no chemical reaction occurs with hydrogen), is illustrated from the work of Den Besten and Selwood[11] for hydrogen on nickel with preadsorbed cyclohexene in Fig. 38.

4. The "Slow" Sorption of H_2 on Ni

All workers in the area of chemisorption are aware that the rapid adsorption of a vapor on the surface of a solid is often followed by a further sorption that may continue for hours or days. The volume of "slow" hydrogen so taken up may be an appreciable fraction of the whole, and the process has received an amount of attention that is, perhaps, out of proportion to its importance. This section will be devoted to the limited amount of magnetization data available for "slow" hydrogen on nickel.[3] Figure 39 shows magnetization–volume isotherms at 303 K for two nickel–kieselguhr catalyst samples. The isotherms were continued beyond the rapid hydrogen uptake for several days until the processes taking place appeared to be virtually complete. (A time scale is shown, but it will be clear that the time consumed during the rapid process is that required for manipulation and for attainment of thermal equilibrium after each hydrogen increment.) It will be noted that in both cases the "slow"-process

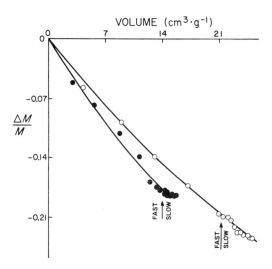

FIG. 39. Magnetization–volume isotherms at room temperature for hydrogen on two commercial nickel catalysts. The data have been extended to include the "slow" sorption occurring over a period of 72 hr.

hydrogen is a moderate fraction of the whole, that the slope of the isotherm changes for one sample, but that it does not change for the other.

The change in rate of adsorption is so abrupt that little difficulty is experienced in determining, in each case, the volume of hydrogen taken up rapidly or slowly, as the case may be. Those quantities related to the rapid process are designated by the subscript 1, and those related to the slow process by the subscript 2.

Let us assume that the rapid sorption of hydrogen takes place on nickel particles of volume v_1, and the slow on those of volume v_2. Then, if the particles are spheres we may relate the volume of the spheres to the hydrogen taken up by each category, as follows:

$$\frac{(N_H)_1}{(N_H)_2} = \frac{(N_p)_1 v_1^{2/3}}{(N_p)_2 v_2^{2/3}}.$$

from which

$$\frac{v_1}{v_2} = \frac{(N_H)_1^{3/2} (N_p)_2^{3/2}}{(N_H)_2^{3/2} (N_p)_1^{3/2}} \qquad (7.3)$$

where $(N_H)_1$, $(N_H)_2$ are the total numbers of hydrogen atoms sorbed on all particles of volumes v_1, v_2, respectively, and $(N_p)_1$, $(N_p)_2$ are the total numbers of particles of volumes v_1, v_2, respectively, in the sample.

Returning now to Eq. (6.20),

$$\Delta M_1/\Delta M_2 = (N_H)_1 (N_p)_1 v_1 / (N_H)_2 (N_p)_2 v_2$$

we may substitute for v_1/v_2 from Eq. (7.3) and obtain

$$\frac{\Delta M_1}{\Delta M_2} = \frac{(N_H)^{5/2}}{(N_H)^{5/2}} \cdot \frac{(N_p)^{3/2}}{(N_p)^{3/2}} \qquad (7.4)$$

This permits us to find $(N_p)_1/(N_p)_2$ from the experimental quantities $\Delta M_1/\Delta M_2$ and $(N_H)_1/(N_H)_2$. The data and the derived ratios are shown in Table VIII.

Table VIII shows that the number $(N_p)_1$ of particles adsorbing hydrogen rapidly is greater than the number $(N_p)_2$ adsorbing slowly, although the ratio is different in the two samples investigated. The volume (v_1) of the individual particles adsorbing rapidly is a little larger in one sample, a little smaller in the other, than the volume (v_2) of those adsorbing slowly. The total volume (V_1) of "rapid" adsorbent is greater in both samples, and the "slow" adsorbent contributes only moderately to the total magnetization in each case. We see, therefore, that a self-consistent set of data emerges from treatment of the "slow" sorption of hydrogen as in no way fundamentally different from the rapid sorption. Equation (6.20) was derived

TABLE VIII

DATA AND DERIVED QUANTITIES RELATIVE TO THE "SLOW"
ADSORPTION OF HYDROGEN ON NICKEL

Quantity[a]	Sample A	Sample B
$\Delta M_1/\Delta M_2$ (obs)	30	5.6
$(N_H)_1/(N_H)_2$ (obs)	12.4	8.2
$(N_p)_1/(N_p)_2 = (\Delta M_2/\Delta M_1)^{2/3}[(N_H)_1/(N_H)_2]^{5/3}$	7.05	39.4
$v_1/v_2 = [(N_H)_1/(N_H)_2]^{3/2}[(N_p)_2/(N_p)_1]^{3/2}$	2.34	0.597
$V_1/V_2 = v_1(N_p)_1/v_2(N_p)_2$	16.5	23.5
$M_1/M_2 = v_1^2(N_p)_1/v_2^2(N_p)_2$	38.6	14.0

[a] ΔM_1, ΔM_2 are the changes of magnetization produced by rapid and slow hydrogen adsorption, respectively. $(N_H)_1$, $(N_H)_2$ are the total numbers of hydrogen atoms adsorbed rapidly and slowly, respectively, on particles of individual volumes (v_1, v_2), of total volume (V_1, V_2), and of magnetizations (M_1, M_2).

on the assumption that ϵ was the same for slow and rapid processes. Certainly no major change in ϵ could occur and still have M_1/M_2 and the other ratios turn out in such a reasonable manner. The implications of these results will be discussed in Chapter VIII.

5. The Systems H₂/Co and H₂/Fe

A very limited number of measurements has been made in these two systems at low fields. Abeledo[12] obtained a magnetization–volume isotherm

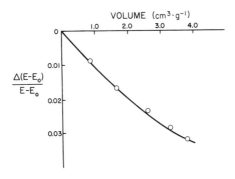

FIG. 40. A plot of fractional secondary emf (proportional to $\Delta M/M$) for hydrogen adsorbed on cobalt–silica at 628 K.

for cobalt supported on silica at 628 K. As shown in Fig. 40 the general trend of results seems to be similar to that observed for hydrogen on nickel at somewhat lower temperature. The very high Curie temperature for cobalt makes it possible to study adsorption processes by the permeameter method at higher temperature than is the case for nickel.

The only nonsaturation study on the H_2/Fe system appears to be that of Artyukh et al.[13] A silica-supported reduced iron sample containing 18% Fe gave $\Delta M/M$ ranging from $+6.3\%$, at 593 K and -4.5% at 77 K. A similar change of sign was found for another sample containing 24% Fe. It is not clear to what degree these results may have been influenced by departures from superparamagnetic behavior, especially at the lower temperatures.

6. Thermal Transients

The abrupt admission of hydrogen to a reduced, evacuated nickel–silica sample always results in magnetization changes as shown in Fig. 41. These consist of a rapid decrease of magnetization followed by a more leisurely partial recovery. This effect is attributed to heating of the nickel particles through liberation of the heat of adsorption, followed by a return to the ambient by the hydrogen-covered particles. There is a possibility, however, that part of the transient effect may be related to hydrogen moving from more to less accessible sites situated on much smaller particles.

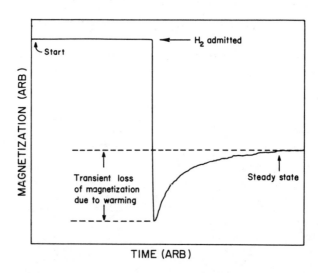

FIG. 41.   Thermal transients observed when hydrogen is abruptly admitted to a nickel–silica sample.

This phenomenon of thermal transients may be used to estimate the heat of adsorption. The change of temperature involved is found by considering that for a sample exhibiting superparamagnetism (the only ones to which the method is applicable) the magnetization must vary inversely as the absolute temperature. A sample of catalyst weighing 5.41 g and containing 52.8% of nickel as metal was abruptly flushed with hydrogen to atmospheric pressure after having previously been evacuated at 673 K. The temperature of adsorption was 293 K. It was found that the magnetization quickly fell until $\Delta M/M = -0.29$, but then came to equilibrium at $\Delta M/M = -0.184$. The total sorption of hydrogen was 45.6 cm³ (STP).

The excess, transient loss of magnetization was 11%, which could have been caused by a temporary rise in temperature of about 31 K. The heat necessary to raise the temperature of 2.85 g nickel metal from 293 to 324 K is about 39 J. This gives a heat of adsorption for hydrogen of about 19 kJ·mol⁻¹. This is, of course, the integrated heat over the whole surface coverage and it is too small by a factor of about 3. We have no accurate knowledge concerning the maximum temperature reached by the nickel. The smaller particles would presumably become hotter than the larger, and we do not know how quickly the heat is distributed to the silica support. Under the circumstances the agreement is about all that could be expected.

A more useful procedure is to use the method for comparing heats of adsorption at different levels of surface coverage. From the change in size of the thermal transients with successive increments of adsorbate it is

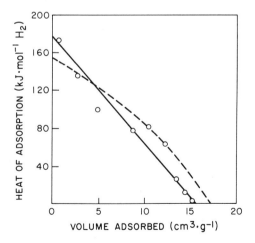

FIG. 42. Incremental heat of adsorption of hydrogen on nickel calculated from thermal transients. The dotted line shows average calorimetric data.

possible to construct a plot of differential heats.[3] Such a plot is shown in Fig. 42. While the absolute accuracy is poor, there is no doubt that the decreasing heat with increasing coverage is clearly reflected by this method which has the unique feature that the nickel is caused to act as its own thermometer.

7. Evaluation of the Low M/M_0 Method

Magnetization measurements made at relatively low values of M/M_0 on superparamagnetic samples give much information, some of which is not obtainable in any other way, but the method has severe limitations. If these limitations are ignored the results obtained may be meaningless. In this section there will be listed the major advantages of the method, the limitations, and some unsolved problems.

The chief advantage for our primary purpose (to gain information about the working of practical catalysts) is that the method is applicable to some very useful preparations of which the familiar nickel–silica is one. Not infrequently, commercial preparations used in large scale hydrogenations prove to be appropriate samples. These samples may be studied over the range of temperature commonly used in actual catalysis, and the method may be used at moderately elevated pressures. It is possible under certain conditions to monitor the magnetization of nickel while it is actually functioning in hydrogenation. Transitory changes lasting only a few seconds may easily be detected. The method is primarily applicable to nickel preparations but it has been demonstrated for cobalt; iron is another possibility. It would appear that various other ferromagnetic or ferrimagnetic samples could be studied in this way. The most useful information gained is the total number of catalyst atoms directly affected by chemisorptive bonding with possible reactant molecules.

A major disadvantage to the low M/M_0 method is that the sample must exhibit superparamagnetism under the actual conditions of temperature and field. An example of the kind of results obtained on a nonsuperparamagnetic sample is shown in Fig. 43. This is for hydrogen on nickel. The adsorption was at room temperature. It will be noted that the magnetic data at 295 K are almost normal, but as the measurement temperature decreases the sample progressively departs more and more from superparamagnetism and the data become meaningless. It is occasionally stated that the low M/M_0 method gives unreliable results. The chief reason for this statement is that the tests for superparamagnetism (see p. 21) have been ignored.

A more serious complication arises from the fact that, especially in certain

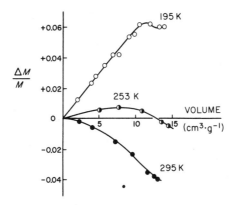

Fig. 43. Magnetization–volume isotherms for hydrogen on nickel–silica at several temperatures, showing deviations from superparamagnetism as the temperature is lowered.

samples, surface coverage of the catalyst by the adsorbate may not be uniform. Martin et al.[14] have examined this possibility in detail. It is, of course, well known that adsorption and reactivity may be different on different exposed crystal faces, although for particles less than 10 nm in diameter it is doubtful if identifiable crystal faces are as significant as random surface irregularity. (The particles we are concerned with are 10 to 100 times smaller than the tips used in field-emission microscopy.) The effects described by Martin et al. are less important for hydrogen than for oxygen and especially for various hydrocarbons. If we assume that hydrogen is uniformly distributed and we find that for H_2 on Ni both high M/M_0 and low M/M_0 methods give the same value for ϵ, it will often be found that the two methods do not give the same value for other molecules. (In Ref. 14 the symbol α is used instead of ϵ as here. $\alpha = 2\epsilon$.) We shall, therefore, defer consideration of this matter until later chapters, except to point out that where any question may arise it would be best to compare the results of both methods at least once on the same system. For such purposes the Lewis development (p. 40) of the gradient magnetometer would appear to offer substantial advantages for both kinds of measurements.

It has already been mentioned that the accessibility of certain adsorbent particles may be less than that of others. If the rate of reaching the surface is significantly less it, of course, complicates the interpretation for both kinds of measurements. On the other hand, it is less likely that such less accessible particles have much influence on a catalysis rate.

One further complication to be mentioned is the subnormal values of ϵ found by Reinen and Selwood[15] from saturation magnetization measurements on Ni–Al$_2$O$_3$ preparations. Such effects could, of course, seriously affect low M/M_0 measurements. This problem has been more extensively investigated by Martin et al.[16] who conclude that, for these and related samples, there must be two different states of chemisorption for H$_2$. In addition to the normal decrease of M, there is a state observed on partially reduced samples. This second state leads to an increase of M. It is difficult to understand this effect, which has not been observed on Ni–SiO$_2$. A possibility is that the highly active hydrogen atoms generated on the surface of the metal may migrate and reduce the NiO adjacent to and, presumably, in contact with the metal. Such a process would certainly have the effect of changing the apparent value of ϵ, if not of actually raising the measured magnetization. A recent paper by Chebotarenko et al.[17] (and available to the writer only through Chemical Abstracts) may offer a clue to the reason for these peculiar effects. Nickel films on silica are found to undergo chemical reaction with the substrate at moderate temperature but only if this contains some excess metal oxide not chemically combined with the silica.

References

1. P. W. Selwood, Rev. Inst. Fr. Pétrole Ann. Combust. Liquides 13, 1656 (1958).
2. I. Den Besten and P. W. Selwood, J. Phys. Chem. 66, 450 (1962).
3. E. L. Lee, J. A. Sabatka, and P. W. Selwood, J. Amer. Chem. Soc. 79, 5391 (1957).
4. J. J. Broeder, L. L. van Reijen, W. M. H. Sachtler, and G. C. A. Schuit, Z. Elektrochem. 60, 838, (1956).
5. J. Geus, A. P. P. Nobel, and P. Zwietering, J. Catal. 1, 8 (1962).
6. J. A. Silvent and P. W. Selwood, J. Amer. Chem. Soc. 83, 1034 (1961).
7. D. O. Hayward and B. M. W. Trapnell, "Chemisorption," 2nd ed. Butterworth, London, 1964.
8. H. Sadek and H. S. Taylor, J. Amer. Chem. Soc. 72, 1168 (1950).
9. G. C. A. Schuit, N. H. de Boer, G. J. H. Dorgelo, and L. L. van Reijen, "Chemisorption," (W. E. Garner, ed.), p. 44. Academic Press, New York, 1957.
10. L. Vaska and P. W. Selwood, J. Amer. Chem. Soc. 80, 1331 (1958).
11. I. Den Besten and P. W. Selwood, J. Catal. 1, 93 (1962).
12. C. R. Abeledo, Doctoral Dissertation, Northwestern University, Evanston, Illinois, 1961.
13. E. N. Artyukh, N. K. Lunev, and M. T. Rusov, Kinet. Catal. 13, 741 (1972).
14. G.-A. Martin, Ph. de Montgolfier, and B. Imelik, Surface Sci. 36, 675 (1973).
15. D. Reinen and P. W. Selwood, J. Catal. 2, 109 (1963).
16. G.-A. Martin, N. Ceaphalan, P. de Montgolfier, and B. Imelik, J. Chim. Phys., (10), 1422 (1973).
17. V. Ya. Chebotarenko, F. M. Matveev, M. Kh. Shorshorov, and B. L. Rudoi, Izv. Akad. Nauk SSSR, Neorg. Mater. 10, 254 (1974); Chem. Abstr. 81, 30028 (1974).

VIII

Hydrogen Bonding on Ni, Ni–Cu, Co, and Fe

1. Conclusions from Magnetic Data

When the comprehensive theory of heterogeneous catalysis is finally formulated it may well be said that an important clue proved to have been hidden in the nickel–hydrogen system. These two familiar elements, whose properties are exhaustively well known, participate mutually in a reversible surface reaction, but then it is found that the hydrogen has undergone a subtle exchange of atomic partners. The system lends itself to many experimental techniques, and it plays a major part in practical catalysis. In this chapter we shall try to present conclusions based primarily on data given in Chapter V and in part in Chapter VII. This first section will be devoted to the facts (as we now see them) that are derived directly from magnetic measurements on small particles.

For nickel and cobalt on silica the change of moment on chemisorption of hydrogen is almost always negative (see p. 56). Those cases in which a positive change of magnetization has been reported have often involved low-field rather than saturation measurements and are readily explained by the diminished anisotropy caused by the adsorption process. A few other cases may involve more than one kind of adsorption or some form of metal–support interaction. In general then formation of the bond must involve one or more of the following steps: (a) donation of an electron to the d band of the metal, (b) pairing of an electron in the localized d level of a nickel atom, (c) some change of spin-orbital coupling, and (d) a diminished ability of the metal atom to participate in the cooperative exchange interaction giving rise to ferromagnetism. This last possibility could even involve negative (antiferromagnetic) interaction with the electron from the hydrogen. In any event, if chemisorption is adsorption associated with more than trivial electronic interaction between adsorbent and adsorbate, there can be no doubt that hydrogen may be chemisorbed on nickel and cobalt.

The magnetization–volume isotherms for H_2 on Ni and on Co are almost always straight lines over the whole range of surface coverage (see p. 56). The exceptions are generally predictable and occur for samples of unusually high specific surface. A few other deviations are referred to below. These results show that for most systems there is no major change in the mechanism of adsorption as coverage rises. The fact that the heat of adsorption may fall by over one order cannot be related to a progressive qualitative change of bond type. Chemisorption thus continues up to relatively high pressures although, as is well known, the quantity of hydrogen adsorbed increases only moderately above about 1 kN·m^{-2}.

The magnetization–volume isotherms are, for the most part, reversible (see p. 60). The important exceptions appear to be mostly for Ni on Al_2O_3 supports—especially if the Ni is only partially reduced to metal. The reversibility must mean that, if there are two or more important modes of bonding, the relative proportions do not change appreciably during adsorption or desorption. The existence of only one kind of chemisorption (in such cases) appears to be the more probable explanation.

The "slow" take up of H_2 by Ni (much studied in the past) has about the same effect (see p. 87) on the magnetization as the "fast" H_2. It appears, therefore, that the bonding mode of the "slow" H_2 is not greatly different from that of the "fast." It may be questioned, incidentally, if the "slow" H_2 is of much significance in catalytic reactions.

Deuterium has about the same effect on Ni as does H_2. No difference in mode of adsorption was expected and no important difference occurs.

The change of Bohr magneton number caused by the adsorption of one atom of hydrogen (ϵ) is most frequently -0.7 for Ni. The closeness, numerically, of ϵ_{Ni} to $\beta(Ni)$ suggests some localization of the Ni–H bond, although this agreement may be fortuitous. However, with only minor reservations, we may say that the magnetic moment of one Ni atom has been effectively destroyed by one H atom. We say "effectively" destroyed because loss of positive exchange interaction on the part of the Ni atom would lower the moment of the metal particle by $\beta(Ni)$, namely, 0.6.

For cobalt (see p. 63) ϵ_{Co} is much less than $\beta(Co)$. If one atom of Co were to lose its ability to contribute collectively to m_p then ϵ_{Co} would be equal, numerically, to $\beta(Co)$, namely, -1.7. But one electron entering the d level of Co would probably not change β(Co) by more than one unit, at most. We can only conclude that the bonding of H_2 on Co is probably related to that on Ni, but some differences remain to be resolved.

For iron (see p. 63) the value of ϵ_{Fe} appears to be $+1.8$, although some uncertainty exists. It has been suggested that atoms in certain surface positions may have lost their cooperative ability and that this is somehow restored by an adsorbed H atom. But much more work on iron is necessary before we may draw any firm conclusions.

For nickel ϵ_{Ni} is independent of particle size over the range for which the magnetic method is applicable (see p. 58). It may be concluded that explanations for the changes in specific catalytic activity and specificity, as a function of particle size, have no explanation in any major changed mode of hydrogen bonding. This does not, of course, necessarily apply to any other adsorbate.

For nickel ϵ_{Ni} rarely changes appreciably with change of adsorption temperature (see p. 58) provided that the temperature is not so low that physical adsorption becomes important. No change of bond type thus occurs for H_2. (Large and important changes occur for many other adsorbates.)

For nickel, so far as it is possible to make a reasonably accurate determination, there is no change of ϵ_{Ni} for measurements at different temperatures (p. 84). A slight change in slope of the magnetization–volume isotherm with temperature is due to changes in M_{sp}. This finding is encouraging because it gives confidence that the all-important measurements at very low temperatures do actually reflect the bonding mode as at higher temperatures.

A change of support, as from SiO_2 to Al_2O_3, causes a change of ϵ_{Ni} under certain circumstances (see p. 61). If this is due to some kind of metal-support interaction and to different modes of binding, then the area is certainly appropriate for further investigation.

A preadsorbed molecule, such as cyclohexene, has no effect on ϵ_{Ni} although the capacity of the surface to take up H_2 is, of course, diminished. This statement is true only if there is no chemical reaction between H_2 and the preadsorbate. This may generally be achieved at moderately low temperature, and it opens a substantial area for the study of various molecules other than, or in conjunction with, H_2 on Ni.

For a 2 Ni–Cu alloy supported on SiO_2, with β(Ni) $= 0.42$, it has been found (see p. 62) that ϵ_{Ni} is -0.37. Other Ni concentrations yield similar results, namely, that ϵ_{Ni} remains approximately equal to the changing value of β(Ni). This result appears to rule out anything but localized decoupling of the Ni atom although, if complete "demetallization" of the affected atom occurs it would have the effect of raising the concentration of

Cu in the particle and hence of lowering the apparent β(Ni) still more. This does not occur.

It might be thought that H_2 chemisorbed on a Ni–Cu alloy surface might be able to migrate in the form of H atoms to neighboring Cu atoms where it could be chemisorbed. If this actually occurred it would explain the magnetic results, but actually the total chemisorption of H_2 is proportional to the surface Ni concentration, as is the heat of adsorption. There is, therefore, apparently no migration, even to adjacent Cu. Why this does not occur is not obvious. Discussion of related matters is given by Sachtler and van der Planck[1] and by Frackiewicz et al.[1a]

2. Related Experimental Data

In this section conclusions derived from experimental methods other than those already described will be presented. Attention will be concentrated on the nickel–hydrogen system in preparations related to supported nickel on silica. Comprehensive reviews of these areas, up to the dates of publication, will be found in the books written by Robertson[2] and edited by Anderson,[3] respectively.

The Ni–H_2 system in which the Ni is in the form of small, supported particles has received massive attention from the point of view of its catalytic activity and its adsorptive properties. But most workers using the powerful physical techniques available have avoided such systems. The reason for this avoidance is because of the belief (which may or may not be true) that supported, practical catalyst systems are necessarily heavily contaminated. It is, consequently, regrettable that no infrared absorption bands[4] are observed for H_2 on Ni although such bands are found for H_2 on Pt and on certain other metals. This negative result has been interpreted as indicating that there is no discrete covalent bond between H and any single Ni metal atom.

Most reported work on adsorption kinetics and equilibria for H_2 on supported Ni, and other metals, was done a number of years ago. A searching review is found in Bond's book[5] and not very much has been added since the date of publication. The earlier work is also reviewed by Hayward and Trapnell.[6] About the only firm conclusions that could be drawn are that the H_2 molecule is dissociated on the metal surface and that at maximum coverage there is approximately one H atom per surface Ni atom.

Adsorption on films is described in comprehensive reviews by Hayward and by Geus (Ref. 3, pp. 225, 327, respectively). From the large volume of work done on films we shall refer to the effect of adsorbed hydrogen on the electrical conductivity and on the magnetization. Both are complicated.

In general the conductivity of a nickel or iron film diminishes on the chemisorption of hydrogen but, not infrequently, the conductivity rises again with increasing surface coverage. The conductivity of a film may depend both on the number of carriers and on the reflection of conduction electrons from the metal surface, and other factors. The general conclusion is that surface hydrides may be formed, repulsive interaction between adsorbed hydrogen atoms may occur, and that this repulsive interaction is less on nickel than on iron. A recent study by Shanabarger[7] of conductivity changes during the desorption of H_2 from Ni films has shown that the rate-determining step involves adsorbed molecular hydrogen. While this is certainly an interesting observation it does not appear to throw much light on our chief problem. The same must be said about the relatively few studies of magnetization changes in films. Where the ratio H_2:Ni is large the effects observed are, at least qualitatively, in agreement with those on small particles, namely, a decrease of M_s.

Results of immediate interest in connection with the interaction of H_2 and Ni are found in a series of papers dealing with nickel saturated electrolytically with H atoms. Baranowski and Smialowski[8] electrolyzed an acidified aqueous solution containing some thiourea as a catalyst poison. The nickel cathode took up hydrogen, and it was found that the hydrogen collected preferentially near the cathodic surface. Subsequently, Bauer and Schmidtbauer[9] reported that the magnetization of nickel so treated had diminished. Apparently a similar effect can occur for iron. Studies of x-ray absorption edges by Faessler and Schmidt[10] on the Ni–H system showed a filling of the 3d band (and of the 4s) of the Ni; Wertheim and Buchanan,[11] by making Mössbauer studies on similarly treated Ni containing [57]Fe were able to confirm that the Ni loses its ferromagnetic properties, that the H is concentrated in the Ni nearest the cathodic surface, and that electrons from the H go, presumably, into the d band. It is of interest also to note that the hydrogen is lost from the Ni at room temperature over a few hours. The total volume of H taken up must, of course, be large compared with chemisorbed layers, and it is probable that surface H remains unless the sample is exposed to air.

From the above results it is tempting to conclude that surface H and bulk H are the same. But the conditions at the surface are not the same. Perhaps it may be concluded, however, that H atoms entering the bulk of Ni are able to destroy the ferromagnetism of the Ni and to do it, presumably, by supplying electrons to enter the d band. Destruction of the atomic moment of a Ni atom would, of course, destroy its ability for cooperative interaction with adjacent atoms. The effect of the H does *not* spread throughout the sample mass. The net result is, therefore, remarkably similar to that occurring during chemisorption. It may be recalled that palladium is able to take up H_2 (without the aid of electrolysis) and that, while ferromagnetism is absent, the moment of one Pd atom is destroyed for each H atom taken up. But even the Pd–H_2 system is much more complicated than previously thought.

3. Theoretical Studies

Recent years have seen a proliferation of theoretical studies on chemisorption. An introduction to this rapidly developing field will be found in the book by Clark.[12] In the following brief account attention will be directed solely to conclusions that may have been reached for H_2 chemisorbed on Ni, Co, or Fe. A major problem in all such work is how to treat the d electrons. It is quite true that ideas having thus far had most influence on heterogeneous catalysis are the qualitative and empirical models developed by authorities having a strong background in that field.

An early attempt by Grimley[13] to calculate interaction energies for several postulated configurations led to the conclusion that a negative hydrogen (H^-) could not be justified and that, if there is a localized orbital with energy lying below the Fermi level, the Ni–H bond is probably covalent. Subsequently, in one of the few studies to take into consideration the change of magnetic moment, Andreev and Shopov[14] and Shopov and Andreev[15] have made molecular orbital calculations that tend to support localized interaction and some residual positive charge on the hydrogen. On the other hand, Horiuti and Toya[16] conceived of two kinds of adsorbed H atoms, with one kind of bond localized in the sense of being directly above one Ni atom but carrying a slight negative charge. The other kind of H atom occupies a position midway between being on the true surface and being interstitial as was described for electrolytic H in Ni in the previous section.

Increasingly sophisticated papers by Schrieffer and Gomer[17] favor the covalent, and necessarily localized, H bond on metals. In a series of papers van der Avoird[18] and Deuss and van der Avoird[19] relates the bonding to the 3d electrons in Ni and provides an explanation for the fact that H bonding on Ni is essentially independent of any activation energy while that on Cu requires a high energy. Further treatment of surface d electrons is given by Johnson,[20] and finally we mention the advanced work of Madhukar[21] who concludes that the H atom retains its full electron spin but that this is coupled antiferromagnetically to an induced spin directly on the metal surface.

It should be emphasized that the work referred to above by no means covers all the important theoretical activity in the area. But it is hoped that a current view has been suggested, namely, that the hydrogen–metal bond is most probably localized and formed through molecular orbitals.

4. Summary of Conclusions

The firm conclusions that may be drawn for the case of H_2 on Ni are few, and they are fewer for H_2 on Co and Fe. If we cannot say very much about the details of surface bonding we can establish a few principles of practical service in the study of adsorbates other than H_2.

Opinion seems to be unanimous (or almost so) that the H_2 molecule is dissociated to atoms on the surface of the metal. The fall of the heat of adsorption with increasing surface coverage is probably due to some residual charge on the hydrogen. This charge is most likely positive. It seems virtually certain that there is a close relation between the number of metal atoms on the surface and the number of H atoms chemisorbed.

There is no major change of bond type in going from the initial to the final state of coverage. Or, at least there is no change in the number and manner in which each metal atom is, in turn, affected. For H_2 on Ni and Co, and perhaps on Fe, there is no major change in bond type, *after* chemisorption has taken place, throughout the temperature range of ~ 2 to ~ 600 K. And there is no major change of bond type with changing temperature of adsorption provided that the temperature is not so low that appreciable physical adsorption takes place. But some small, progressive changes may occur in the adsorbent on repeated adsorption–desorption cycles. Otherwise all the changes are reversible. (It must be emphasized again that this is not necessarily true of adsorbates other than H_2.)

In general, there is no major change of bond type with changing particle size, but there appear to be some exceptions and, again, some doubts concerning other adsorbates.

The concept of surface molecular orbital bonding appears to be gaining strength. But, if this is correct, it is not clear why no metal–H infrared stretching frequency is observed.

There is strong but not conclusive evidence that the magnetic moment of Ni is destroyed (presumably by electron pairing). If this occurs it would, of course, preclude cooperative interaction with other metal atoms in the particle.

In conclusion about all we can say about the theory of surface bonding is that the theoretical methods all indicate a complex interaction. In the words of Clark[12]: "The fortress of chemisorption continues to guard successfully many secrets."

References

1. W. M. H. Sachtler and P. van der Planck, *Surface Sci.* **18**, 62 (1969).
1a. A. Frackiewicz, Z. Karpinski, A. Leszczynski, and W. Palczewska, "Proceedings of the Fifth International Congress of Catalysis, 1972," p. 635. North-Holland, Amsterdam, 1973.
2. A. J. B. Robertson, "Catalysis of Gas Reactions by Metals." Logos Press, London, 1970.
3. J. R. Anderson, ed., "Chemisorption and Reactions on Metal Films," Vol. I. Academic Press, New York, 1971.
4. W. A. Pliskin and R. P. Eischens, *Z. Phys. Chem. F* **24**, 11 (1960).
5. G. C. Bond, "Catalysis by Metals." Academic Press, New York, 1962.
6. D. O. Hayward and B. M. W. Trapnell, "Chemisorption," 2nd ed. Butterworth, London, 1964.
7. M. R. Shanabarger, *Solid State Commun.* **14**, 1015 (1974).
8. B. Baranowski and M. Smialowski, *J. Phys. Chem. Solids* **12**, 206 (1959).
9. H. J. Bauer and E. Schmidtbauer, *Z. Phys.* **164**, 367 (1961).
10. A. Faessler and R. Schmidt, *Z. Phys.* **19**, 10 (1966).
11. G. K. Wertheim and D. Buchanan, *Phys. Lett.* **21**, 255 (1966).
12. A. Clark, "The Chemisorptive Bond." Academic Press, New York, 1974.
13. T. B. Grimley, *in* "Chemisorption" (W. E. Garner, ed.). Academic Press, New York, 1957.
14. A. Andreev and D. Shopov, *C. R. Bulgarian Acad. Sci.* **22**, 887, (1969).
15. D. Shopov and A. Andreev, *J. Catal.* **13**, 123 (1969).
16. J. Horiuti and T. Toya, "Solid State Surface Science" (M. Green, ed.), Vol. I. Dekker, New York, 1969.
17. J. R. Schrieffer and R. Gomer, *Surface Sci.* **25**, 315 (1971).
18. A. van der Avoird, *Surface Sci.* **18**, 159 (1969).
19. H. Deuss and A. van der Avoird, *Phys. Rev. B* **8**, 2441 (1973).
20. O. Johnson, *J. Catal.* **28**, 503 (1973).
21. A. Madhukar, *Phys. Rev. B* **8**, 4458 (1973).

IX

Determination of Bond Number, O_2, CO, CO_2, H_2S, $(CH_3)_2S$, and N_2

1. Bond Number

This chapter and Chapters X and XI will describe how magnetic methods may give information concerning the mode of bonding for molecules more complex than molecular hydrogen on nickel and, to a much more limited degree, on cobalt. There appear to be no experimental data for iron as adsorbent in this area. The importance of obtaining this kind of information is obvious. When it becomes possible to describe in detail how a molecule of, say, benzene is chemisorbed on nickel metal we shall have taken a major step in the elusive understanding of heterogeneous catalytic hydrogenation. Fortunately there are now increasingly diverse methods with which the conclusions derived from magnetic measurements may be compared and it will be shown that, in more than a few cases, agreement is gratifying.

The magnetic method for determining bond number consists of, in brief, comparing the change of magnetic moment produced by one molecule of an adsorbate (X) with that produced by one atom of hydrogen on the same adsorbent. We define the bond number of any such molecule, on Ni, as $\zeta_{Ni}(X) = \epsilon_{Ni}(X)/\epsilon_{Ni}(H)$ where $\epsilon_{Ni}(X)$ is given (for nickel adsorbent) by Eq. (5.3) with the exception that the number of hydrogen *atoms* is replaced by the number of X *molecules*. Thus defined, the bond number is seen to be very nearly the number of adsorbent atoms that, for one reason or another, have lost their ability to contribute to the magnetic moment of the adsorbent particle, per each molecule adsorbed. This, of course, is based on the assumptions that H_2 is chemisorbed over the range of experimental conditions by a dissociative mechanism leading to Ni–H bonds, that the H atom is attached to one, and only one, metal atom, and that the nature of the bond does not change with changing surface coverage or, within limits, of adsorption or measurement temperature.

The definition of bond number given in the preceding paragraph shows that experimental determinations require measurement of M_0 and M_0'. Fortunately, these data are now available for a substantial number of adsorbates. But with some restrictions, the low-field permeameter method yields satisfactory estimates of ζ, without the rather tedious recourse to very low temperatures and high fields. Equation (6.5) shows that $\Delta M/M$ is proportional to $\Delta M_s/M_s$ provided that superparamagnetism is exhibited, but it is obvious that comparison of isotherms produced by one adsorbate and another is a valid procedure only if the adsorbent remains in the same condition with respect to volume (and saturation magnetization), temperature, and distribution of particle sizes. These conditions may be met approximately if the identical sample is completely reduced and evacuated, used to obtain a hydrogen isotherm at the desired temperature, then evacuated again with no significant nickel particle growth prior to obtaining of data for the second isotherm. Actually, these conditions offer no more difficulty than many standard procedures in physical chemistry.

Having obtained our two isotherms we may relate the slopes to the number of bonds formed thus, if 1 cm³ (STP) of vapor X_2 lowers the magnetization just twice as much as 1 cm³ of H_2, then molecule X_2 is held to the nickel by four bonds or, at least, four nickel atoms no longer contribute to the magnetic moment of the particle. The degree to which we may rely on this simple procedure will be illustrated for several cases in what follows.

2. Oxygen

If the concept of chemisorbed oxygen has any physical meaning it is clear that oxygen yields nothing to any other molecule in complexity. Furthermore, the direct addition of oxygen to a reduced supported nickel sample often results in the partial, or complete, conversion of the nickel to nickel oxide.

We shall start with the magnetic saturation studies of Martin et al.[1-3] which is the reverse of the chronological order. Measurements were made on nickel,[1,2] cobalt,[1] and nickel–copper[3] alloy and, in each case, a comparison was made with H_2 as adsorbate. Details of O_2 admission are not given but this was, presumably, at low pressure and near room temperature. The results obtained at 4.2 K show $\epsilon_{Ni}(H) = -0.65$ and $\epsilon_{Ni}(O_2) = -1.5$, with straight-line magnetization isotherms and no appreciable variation of ϵ over a fairly wide range of particle sizes. (The results, incidentally, show evidence of preferential adsorption of O_2 on smaller particles of Ni.) Cobalt isotherms are also straight lines with $\epsilon_{Co}(O_2) = -1.3$. Examples of the isotherms are shown in Fig. 44. On a 2Ni–Cu alloy $\epsilon_{Ni-Cu}(O_2) \simeq -0.2$ which is not unreasonable, but the volume of O_2 adsorbate was considerably larger than that of H_2 on the same sample, thus suggesting reaction below the surface.

Results of relatively low-field studies on O_2 adsorbed on Ni have been reported by Broeder et al.[4] who admitted the O_2, as such, at 77 K. This procedure minimizes oxidation of the metal below the surface. The isotherm at room temperature is a straight line with negative slope. (In an earlier paper the writer reported a positive effect for oxygen. This has since been shown to be almost certainly due to failure to consider anisotropy effects at the conditions used.) A typical isotherm obtained by Leak,[5] together with one for H_2 on the same sample, both at room temperature, is shown in Fig. 45. This is in substantial agreement with Broeder et al.[4] (It may be added that the problems encountered through the use of pure O_2 as adsorbate at, or near, room temperature may be avoided by adding the O_2 in about 0.2% mixture in helium. Alternatively, nitrous oxide, that decomposes instantly to N_2 and O_2 over activated nickel at room temperature, may be used as a source of O_2. The possibility that N_2 is chemisorbed on Ni

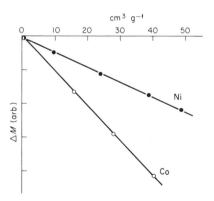

FIG. 44. Saturation magnetization–volume isotherms for oxygen on nickel and on cobalt (after Dalmon et al., Ref. 1 and Martin et al., Ref. 2).

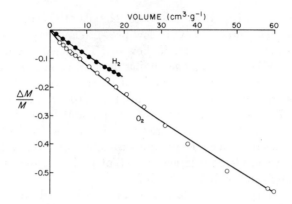

FIG. 45.   Low-field isotherms for hydrogen and for oxygen on nickel–kieselguhr, at room temperature.

will be discussed later.) Still other low-field results in agreement with Broeder[4] and with Leak[5] are reported by Geus et al.,[6] and a comprehensive and very careful study by Geus and Nobel[7] clears up various uncertainties in this difficult area. Their conclusions will be summarized in the following paragraph.

There is a difference in the action of molecular O_2 and oxygen derived from N_2O. The former almost always leads to successive oxidation of the metal particles, and the latter to a homogeneous distribution. In a superparamagnetic sample oxygen from N_2O decomposition has a larger effect on the magnetization and the effect is a decrease. It appears from this work and from the other magnetic data that a molecule of O_2 has about the same effect on the nickel particle as a molecule of H_2, which is to say that one Ni atom is bonded for each oxygen atom adsorbed. (This is, of course, not to imply that the mechanism of bonding is the same.) For samples not exhibiting superparamagnetic behavior, under the experimental conditions, positive changes in magnetization may be observed. The distribution of oxygen on the surface may determine the sign of the magnetization change. There may be minor effects of weak exchange coupling between surface atoms and atoms in the particle core. And finally, there is substantial evidence that more than one mode of chemisorption may occur. There have been extensive low-energy electron diffraction studies of the Ni–O_2 system, on single crystals. The results are not directly comparable with those reported here, but appear to be not inconsistent with the results of direct molecular O_2 adsorption. The same may be said of electrical conductivity work on Ni films.

3. Carbon Monoxide

With the exception of hydrogen and possibly of ethylene, no adsorbate on metals has been more thoroughly investigated than carbon monoxide. But the infrared adsorption method has scored a notable success in this area and, as we shall see below, the magnetic results are at least in qualitative agreement. We shall first review the earlier low-field magnetic investigations.

A magnetization–volume isotherm[8] for carbon monoxide adsorbed on nickel–silica at 298 K is shown in Fig. 46. The isotherm for hydrogen under the same conditions on the same sample is also shown. As the pressure of carbon monoxide is increased, a large additional sorption occurs, but the magnetization isotherm suffers a rather abrupt change of slope, becoming much more nearly parallel to the volume axis. It is well known that nickel tetracarbonyl is formed under these conditions, but the rate of formation at room temperature is so slow as not to interfere seriously with our interpretation of the data. The nickel particle diameters were in the range of 2.5 nm.

Up to the point of change of slope the carbon monoxide cannot be evacuated at room temperature; evacuation at elevated temperature yields a substantial fraction of carbon dioxide formed, presumably, by disproportionation. But at higher surface coverages it is possible to evacuate some of the carbon monoxide as such at room temperature, and substantially all of it if the temperature is raised to 423 K. These changes will

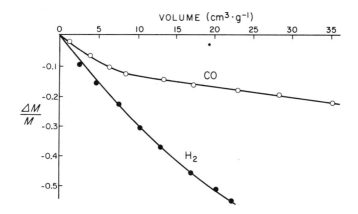

Fig. 46. Low-field isotherms for hydrogen and for carbon monoxide on nickel–silica, at room temperature.

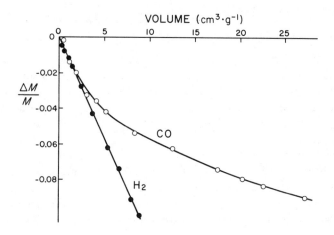

FIG. 47.   Low-field isotherms for hydrogen and for carbon monoxide on nickel–
kieselguhr at room temperature. The nickel particle volumes were approximately 15
times greater than those used for the results shown in Fig. 46.

perhaps be clearer if presented in the form of an example. A nickel–
kieselguhr at 298 K took up a total of 29.4 cm³ CO/g Ni, of which 16.2
cm³ was beyond the change of magnetization slope. Then, on evacuation
5.0 cm³ was desorbed at 298 K and 9.5 cm³ more at 423 K, this latter fraction being about 95% carbon monoxide and 4% carbon dioxide. By evacuation at 673 K it proved possible to recover nearly all of the carbon but
most of this was in the form of carbon dioxide. During these changes the
magnetization, as measured at 298 K, rose to about 95% of its initial value.

There is evidence in the literature that the chemisorption mechanism of
CO on Ni is dependent on particle size. Figure 47 shows an isotherm obtained on a nickel–kieselguhr that had been briefly sintered at 873 K to
cause particle growth. The particle diameter for this sample was in the
range 6.4 nm, which brings it to the upper limit of applicability of the
low-frequency ac permeameter. In spite of this drawback it appears that
the initial slope of the isotherm is the same as that for hydrogen, in contrast to the results on smaller particles shown in Fig. 46. These results are
in qualitative agreement with those reported by Geus et al.[6] for supported
Ni in the particle diameter range of 4 nm.

Our conclusions with respect to carbon monoxide, based solely on the
above magnetic studies, are as follows.

1. On small nickel particles the initial stage of chemisorption is probably
mostly linear, O=C=Ni.

2. At higher coverages it appears that two (or more) carbon monoxide molecules are adsorbed on one nickel atom in a fashion such as

$$
\underset{C}{\overset{O}{\diagdown}}\underset{Ni}{\diagup}\overset{O}{\overset{C}{\diagup}} \quad \text{or} \quad \underset{C}{\overset{O}{\diagdown}}\underset{Ni}{\diagup}\underset{C}{\overset{O}{\|}}\underset{Ni}{\diagdown}\overset{O}{\overset{C}{\diagup}}
$$

3. There is a strong dependence of the mode of adsorption on nickel particle size. On larger particles the initial mode of adsorption appears to be bridged, that is to say, two nickel atoms are involved for every carbon monoxide molecule adsorbed.

$$
\underset{Ni \quad Ni}{\overset{O}{\overset{\|}{C}}}
$$

4. When two (or possibly more) carbon monoxide molecules are attached to one nickel atom, desorption of the monoxide as such becomes quite easy.

5. Carbon dioxide is probably formed upon desorption from nickel atoms bonded to only one carbon monoxide molecule.

6. The low chemisorption of carbon dioxide to be described in the following section suggests that disproportionation of carbon monoxide does not occur until desorption is attempted.

The very extensive infrared studies on the Ni–CO system are reviewed (to 1965) by Little.[9] Unfortunately, the infrared results are not quite so nearly unambiguous as for many other systems. Nevertheless, the important work of Eischens et al.[10] and of Yates and Garland[11] show that the frequencies of bridging carbonyl appear first and that linear bonding appears at higher coverage. Lacking information on the particle diameters involved we can say that, if the nickel particles used in the infrared work were comparable to those used in obtaining the magnetic data of Fig. 46, then the agreement between the two methods is satisfactory.

Some very recent magnetic saturation work at relatively low surface coverage by Martin and associates (personal communication) has indicated that at 473 K the bond number for CO on pure Ni is 3.8. Parallel infrared experiments (by M. Primet) showed that no linear or bridged species was observed under these conditions, but that a hitherto unobserved band at 1840 cm^{-1} could be assigned to CO bonded to three or four Ni atoms. The agreement here is, therefore, excellent. Equally interesting is the report by Dalmon et al.[11a] involving parallel magnetic and infrared studies of CO adsorbed on Ni–Cu alloys. For adsorption at low coverage at room temperature on pure Ni the bond number is 1.8 indicating chiefly a bridged complex, but as the Cu content increases the bond number falls to 1.0 indicating a linear species. These and other conclusions from the magnetic data are in agreement with those from infrared, on the same samples.

4. Carbon Dioxide

Magnetization studies on the Ni–CO₂ system appear to be limited to those described by Den Besten *et al.*[8] Figure 48 shows low-field data obtained on a nickel–kieselguhr sample at 298 K. At a pressure of 31 kN·m⁻² the total sorption was 9.0 cm³ (STP) g⁻¹ Ni but evacuation at room temperature removed all but 20% of this without change of magnetization. An additional 10% may be removed by evacuation at higher temperature. The total volume of truly chemisorbed CO₂ on Ni is, therefore, quite small although no smaller than that of ethane under the same conditions. The initial slope of the CO₂ isotherm is moderately greater than that of H₂.

Our conclusion with respect to carbon dioxide is, therefore, that the molecule is held by at least two bonds and that for some reason which is obscure the maximum surface coverage is approximately one-eighth of that possible with hydrogen. In the case of ethane the demonstrated occurrence of dissociative adsorption, and the surface-covering ability of any molecule requiring two adjacent sites, makes it easy to understand why the apparent coverage is small. With carbon dioxide there is no obvious reason why the coverage should be less than half that of carbon monoxide. Actually it is

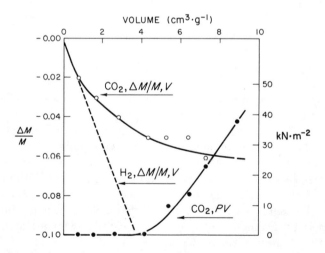

Fig. 48. Low-field magnetization–volume and pressure–volume isotherms for carbon dioxide on nickel–kieselguhr at room temperature.

barely 6% as great. We are forced to take recourse in the old idea that there are sites of greater and lesser activity on nickel. Only the sites of greatest activity are able to chemisorb carbon dioxide. It will be shown below that adsorbed inert gases such as krypton may have a measurable effect on the magnetization of nickel (probably owing to polarization), and it might be thought that a similar effect could occur with carbon dioxide, but the effect of a molecule of krypton is only 16% of that of a molecule of carbon dioxide.

The infrared absorption spectrum of CO_2 on Ni-SiO_2 has been reported by Eischens and Pliskin.[12] The bands found are characteristic of the carboxylate ion and suggest the following mode:

$$
\begin{array}{c}
O \diagdown \diagup O^- \\
C \\
| \\
Ni
\end{array}
$$

At moderately elevated temperatures of adsorption there was evidence of carbon monoxide, and this suggests that any oxygen liberated may have entered the metal. Eischens and Pliskin also studied physically adsorbed CO_2 and the results seem to preclude this possibility on the nickel samples investigated. The heterogeneity view suggested here appears to be supported also by tracer studies of Kobayashi and Hirota.[13]

5. Hydrogen Sulfide and Dimethyl Sulfide

Figure 49 shows low-field magnetization–volume isotherms obtained by Den Besten and Selwood[14] for hydrogen and hydrogen sulfide on a commercial nickel–kieselguhr reduced and evacuated in the usual way. The temperature of adsorption and of measurement was 298 K. This pair of isotherms, typical of many, shows that the slope of the hydrogen sulfide isotherm is, within ±5%, twice that of molecular hydrogen. The implication is that hydrogen sulfide is, under these conditions, dissociatively adsorbed thus

$$
\begin{array}{ccc}
H & & H \\
| & \diagup S \diagdown & | \\
Ni & Ni Ni & Ni
\end{array}
$$

(there being, of course, no implication as to whether hydrogen or sulfur is attached on or between the nickel atoms). Isotherms obtained at 273 and 338 K give almost identical results. We may, therefore, conclude on the basis of the low-field magnetic data that dissociation of the hydrogen

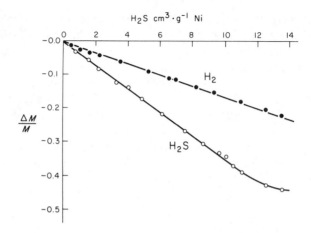

Fig. 49.   Low-field isotherms for hydrogen and for hydrogen sulfide on nickel–kieselguhr at room temperature.

sulfide is complete, and that there is no change of bond type over the temperature range investigated.

Support for the view stated above is available from at least two sources. First, even up to quite appreciable pressures the only gas found in the free space over the catalyst is hydrogen. Furthermore, provided that surface coverage is moderate it is possible at 633 K to evacuate nearly all the hydrogen as molecular hydrogen, but none as hydrogen sulfide. Thus, after the adsorption of 1.43 cm³ H₂S/g Ni at 298 K, one may remove 1.36 cm³ H₂ (as such) by evacuation at 633 K. It is difficult to see how these results could be obtained if complete dissociative adsorption had not occurred.

The most convincing agreement is, however, that provided by Kemball[15] who shows that, in the temperature region covered, two hydrogen atoms, per molecule of hydrogen sulfide adsorbed, are readily exchange for deuterium.

Saturation magnetization measurements on the H₂S–Ni system have been made by Martin and Imelik[16] who show that the bond number, $\zeta_{Ni}(H_2S)$, is about 2.5 for adsorption slightly below room temperature. For higher temperatures of adsorption the bond number rises rapidly, indicating progressive dissociation. These results are, therefore, in qualitative agreement with those obtained by the low-field method.

At higher surface coverages of hydrogen sulfide, more complicated effects occur. One of these is that as the pressure over the sample becomes appreciable, it may be observed that each additional increment of adsorbed hydrogen sulfide results in the rather slow liberation of hydrogen, so that some increase of hydrogen pressure actually occurs. This is in sharp contrast to the slow disappearance of hydrogen from the gas phase when, under similar conditions, hydrogen itself is the adsorbate. This effect is probably due to progressive dissociation of chemisorbed hydrogen sulfide molecules on a surface which, because it is already nearly covered with hydrogen, can accept no more hydrogen, but the more tightly bound sulfur is, nevertheless, readily accepted.

It is also found that after surface coverage is virtually complete, it becomes impossible to remove more than about 70% of the hydrogen either by evacuation at 633 K or by exchange with deuterium. The reason for this is not clear, but it may be related to blocking of the surface by sulfur which retards egress of interstitial hydrogen. In this connection it is found that addition of hydrogen sulfide to a surface already covered by hydrogen results in liberation of some hydrogen from the surface. It will also be noted that the cumulative loss of magnetization caused by hydrogen sulfide is nearly twice that produced by hydrogen as surface coverage appears to be nearing completion. If our views concerning the ability of hydrogen to destroy the magnetization of each surface nickel atom are correct, then it is obvious that the sulfur from hydrogen sulfide must involve more than the surface layer of nickel atoms. This view is also consistent with Kemball's findings, the only important difference being that changes occur somewhat more readily and at lower temperatures on evaporated films. Blyholder and Bowen[17] have reported, however, that the chemisorption of H_2S on Ni at room temperature is too small to detect by infrared.

It may be noted that Richardson[18] has used an adaptation of the permeameter method to monitor the progressive poisoning of a Ni catalyst by H_2S during the course of a hydrogenation reaction.

Dimethyl sulfide as a catalyst poison is even more notorious than hydrogen sulfide. Magnetization–volume isotherms for this adsorbate have been obtained by Den Besten and Selwood[14], as shown in Fig. 50 for the temperatures 298 and 393 K. The initial slope at room temperature is only slightly greater than that for hydrogen. This suggests bonding to two nickel atoms as follows.

$$\begin{array}{c} H_3C \diagdown \quad \diagup CH_3 \\ S \\ \diagup \diagdown \\ Ni \qquad Ni \end{array}$$

which is in a manner resembling that suggested by Kemball for hydrogen sulfide on nickel films in the neighborhood of 193 K.

The adsorption mechanism of dimethyl sulfide is quite sensitive to temperature. At 393 K the magnetic data indicate extensive dissociation—

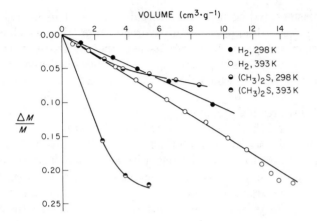

F IG. 50. Low-field isotherms for hydrogen and for dimethyl sulfide on nickel–kieselguhr at 298 and 393 K.

the initial slope being consistent with at least 10 bonds being formed per molecule adsorbed. This conclusion is confirmed by the nature of the desorption products obtained as the temperature is gradually raised to 673 K. These products include methane, ethane, and hydrogen, but no sulfur compound. Our view of catalyst poisoning by Lewis bases such as hydrogen sulfide and dimethyl sulfide is, therefore, the classical one of atoms, such as sulfur, bonded to the catalyst sites more strongly than hydrogen, at at least more strongly than adsorbed hydrogen in a reactive state. The greater efficiency of dimethyl sulfide as a catalyst poison appears to be related to its greater covering power per molecule adsorbed.

There appear to be no other studies with which to compare the magnetic data on dimethyl sulfide, excepting that on palladium, to which further reference will be made later. But Blyholder and Bowen[17] have studied the infrared absorption spectrum of diethyl sulfide and report extensive dissociation with the formation of Ni–C bonds. This is in agreement with the magnetic results on (CH₃)₂S.

The reader may wonder why the chemisorption of water has not been included in this chapter. The reason is that the extraordinary affinity of silica and other catalyst supports (pretreated at elevated temperatures) makes interpretation difficult or impossible.

6. Nitrogen

The chemisorption of nitrogen on iron is a phenomenon of prime importance in the synthesis of ammonia. It is to be hoped that in due course the magnetic method will help to throw some light on this obscure process. For the present we shall have to confine our attention to nitrogen on nickel.

Over the years there have been several claims that nitrogen may be chemisorbed on nickel. The two studies of direct concern to us are those of Schuit and de Boer[19] and of Kokes and Emmett,[20] the former having used nickel–silica preparations quite similar to those used in many of the magnetic studies already described. Schuit and de Boer concluded that nitrogen may be chemisorbed on nickel because, primarily, the volume adsorbed at 195 K was found to depend on the amount of nickel present and not on the total surface as determined by nitrogen adsorption at 77 K. This work carries the implication that there is no chemisorption of nitrogen on nickel at 77 K.

Kokes and Emmett, on the other hand, reached the conclusion that nitrogen may be chemisorbed on nickel at 77 K. (If this is true, it tends to invalidate the Schuit and de Boer thesis.) The method used by Kokes and Emmett was to compare the total sorption of nitrogen at 77 K before and after evacuation at 195 K. The difference was attributed to chemisorbed nitrogen. There was, however, evidence that some of the nitrogen thought to be chemisorbed at 77 K was removed by evacuation at 195 K, and this introduces a degree of uncertainty concerning the actual volume of nitrogen which may be said to be chemisorbed.

There seems little reason to doubt the experimental facts presented by either group of investigators. If the effect described by Schuit and de Boer does not represent chemisorption, then we have no ready explanation for it. As for the conclusions reached by Kokes and Emmett, it should be pointed out that the method actually provides evidence which can at best be described as circumstantial. This is the kind of evidence that would provide useful confirmation provided that some direct evidence concerning electronic interaction were at hand. In the absence of such direct evidence we must refer once more to Barrer's comment[21] concerning the erroneous conclusions which may be based on sorption energy considerations in certain adsorbents of which silica gel is one. Whether this has any applicability to the unsupported nickel used by Kokes and Emmett we are not in a position to say.

Nitrogen sorbed on nickel has also been shown to affect the electrical conductivity and to cause the appearance of a surface dipole, but the effects are no larger than those produced by xenon and cannot be considered proof of chemisorption at the present state of our understanding of these effects.

The effect of adsorbed nitrogen on the magnetization of silica-supported nickel has been investigated by the writer.[22] Nitrogen does, indeed, cause some loss of magnetization. The effect, which appears to reach a maximum

in the neighborhood of 223 K is quite small, amounting to only a few per-
cent of the effect observed with an equal volume of adsorbed hydrogen. A
magnetization-volume isotherm for nitrogen on nickel–silica at 195 K is
shown in Fig. 51. The isotherm was found to be reversible—only a trace
of nitrogen remaining adsorbed after evacuation at 195 K—and all the
magnetization was recovered.

These results argue against any true chemisorption on the part of nitrogen under these conditions. It may be thought that perhaps the surface
was already covered by preadsorbed nitrogen throughout the whole experiment. But the gas purification methods used make this contingency
improbable, if not completely impossible. Even more convincing evidence
is obtained by examining the effect of adsorbed inert gases. Helium is not
adsorbed and shows no magnetic effect, even up to 140 atm.[23] Argon shows
an effect comparable in magnitude with that of nitrogen; krypton at 195
K causes a loss of magnetization which is about 20% of that caused by
hydrogen. The reason for these magnetic effects of adsorbed inert gases
(and possibly also of nitrogen) may lie in the polarization[24] of the adsorbate
molecule which, in turn, causes some slight change of electron density in
several bands. This view is supported by the fact that krypton, the most
polarizable molecule of the group, showed the largest effect.

The evidence presented in preceding paragraphs is by no means the whole
story. The infrared absorption spectrum appears to show clear evidence of
chemisorbed linear nitrogen, N≡N, on nickel although the quantity of N₂
taken up is small. Merten and Eischens[25] have made simultaneous magnetic
and infrared measurements, and they find that while the maximum
possible coverage at 298 K with N₂ is only 10% that with H₂, and the
change of magnetization is correspondingly small, yet the slope of the

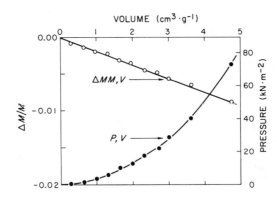

FIG. 51. Low-field isotherm for nitrogen on nickel–kieselguhr at 195 K.

isotherm for N_2 is almost the same as that for H_2. Approximately the same results have been obtained by Wösten et al.,[26] although Van Hardeveld and Van Montfoort[27] advance the idea that only very small particles participate in the effect, state that it is actually a physical adsorption, and that the effects are due to a special kind of physical adsorption on certain faces. The problem is further discussed by Nieuwenhuys and Sachtler.[28]

In conclusion we may say that if chemisorption of N_2 actually occurs on Ni it involves only a small fraction of the surface. But then it is difficult to deny that argon and especially krypton are also chemisorbed. As yet there do not appear to be any magnetic measurements on the $Fe-N_2$ system.

References

1. J. A. Dalmon, G.-A. Martin, and B. Imelik, Coll. Intern. CNRS, 201 Thermochimie, Marseille, 1971, p. 593.
2. G.-A. Martin, Ph. de Montgolfier, and B. Imelik, Surface Sci. 36, 675 (1973).
3. J. A. Dalmon, G.-A. Martin, and B. Imelik, Surface Sci. 41, 587 (1974).
4. J. J. Broeder, L. L. van Reijen, and A. R. Korswagen, J. Chim. Phys. 54, 37 (1954).
5. R. J. Leak, J. Phys. Chem. 64, 1114 (1960).
6. J. W. Geus, A. P. P. Nobel, and P. Zwietering, J. Catal. 1, 8 (1962).
7. J. W. Geus and A. P. P. Nobel, J. Catal. 6, 108 (1966).
8. I. E. Den Besten, P. G. Fox, and P. W. Selwood, J. Phys. Chem. 66, 450, (1962).
9. L. H. Little, "Infrared Spectra of Adsorbed Species." Academic Press, New York, 1966.
10. R. P. Eischens, S. A. Francis, and W. A. Pliskin, J. Phys. Chem. 60, 194 (1956).
11. J. T. Yates and C. W. Garland, J. Phys. Chem. 65, 617 (1961).
11a. J.-A. Dalmon, M. Primet, G.-A. Martin, and B. Imelik, Surface Sci. (to be published).
12. R. P. Eischens and W. A. Pliskin, Advan. Catal. 9, 662 (1957).
13. Y. Kobayashi and K. Hirota, Bull. Chem. Soc. Japan 39, 453 (1966).
14. I. E. Den Besten and P. W. Selwood, J. Catal. 1, 93 (1962).
15. C. Kemball, Actes Congr. Intern. Catalyse, 2e, Paris, 1960 2, 1811 (1961).
16. G.-A. Martin and B. Imelik, Surface Sci. 42, 157 (1974).
17. G. Blyholder and D. O. Bowen, J. Phys. Chem. 66, 1288 (1962).
18. J. T. Richardson, J. Catal. 21, 130 (1971).
19. G. C. A. Schuit and N. H. de Boer, J. Chim. Phys. 51, 482 (1954).
20. R. J. Kokes and P. H. Emmett, J. Amer. Chem. Soc. 82, 1037 (1960).
21. R. M. Barrer, in "Chemisorption" (W. E. Garner, ed.), p. 91. Academic Press, New York, 1957.
22. P. W. Selwood, J. Amer. Chem. Soc. 80, 4198 (1958).
23. L. Vaska and P. W. Selwood, J. Amer. Chem. Soc. 80, 1331 (1958).
24. R. A. Pierotti and G. D. Halsey, Jr., J. Phys. Chem. 63, 680 (1959).
25. F. P. Merten and R. P. Eischens, in "The Structure and Chemistry of Solid

Surfaces" (G. A. Somorjai, ed.), p. 53-1 (Proc. Intern. Mater. Symp., 4th, Berkeley, 1968). Wiley, New York, 1968.

26. W. J. Wosten, Th. J. Osinga, and B. G. Luisen, *in* "The Structure and Chemistry of Solid Surfaces" (G. A. Somojai, ed.), p. 54-1 (Proc. Intern. Mater. Symp., 4th, Berkeley, 1968). Wiley, New York, 1968.

27. R. Van Hardeveld and A. Van Montfoort, *Surface Sci.* **4**, 396 (1966).

28. B. F. Nieuwenhuys and W. M. H. Sachtler, *Surface Sci.* **34**, 317 (1973).

X

Ethane, Ethylene, and Acetylene

1. Ethane

It is sometimes stated that ethane is not chemisorbed to any extent on nickel, or that, together with iron and cobalt, nickel is almost completely inactive toward ethane sorption. These statements are quite misleading—Trapnell's[1] data show a small but definite chemisorption, and the exchange results of Kemball and others, summarized by Anderson,[2] show clearly that ethane may undergo exchange over nickel at moderate temperatures. It is difficult to see how ethane can exchange any hydrogen for deuterium unless the molecule of ethane is first dissociatively adsorbed. This view that ethane is indeed chemisorbed on nickel receives further confirmation from the infrared absorption spectra obtained by Eischens and Pliskin[3] who show that on a bare nickel surface at 308 K ethane gives a spectrum similar to that of ethylene on bare nickel.

Figure 52 shows a low-field magnetization–volume isotherm with adsorption and measurement at 300 K for ethane on a bare nickel–kieselguhr surface, and also on the same surface almost completely covered by preadsorbed hydrogen.[4] There is appreciable van der Waals adsorption at this temperature, but the data clearly show a small but readily measurable chemisorption of ethane on nickel under these conditions. A rough estimate suggests a minimum of eight bonds (erroneously given as four in the original paper) for the average ethane molecule. The ethane on the bare surface appears, obviously, to be dissociatively adsorbed, and this is in agreement with the infrared results.[3] However, Anderson and Kemball[5]

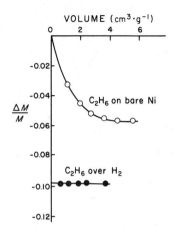

FIG. 52. Low-field magnetization–volume isotherms for ethane and for ethane over preadsorbed hydrogen on nickel–kieselguhr, all at room temperature.

have shown that the mode of adsorption of ethane is very much dependent on the experimental conditions. It is, therefore, important to bear in mind that on certain preparations the low-field method may give erroneous results owing to preferential adsorption on some metal particles (p. 92). Fortunately, a careful magnetic saturation study of ethane and other hydrocarbons has been made by Martin and Imelik.[6] Their nickel–silica adsorbents were prepared by impregnation, and reduced at the high temperature of 893 K. Samples were analyzed chemically for nickel metal which proved to be 100% of the nickel present. Gases were adsorbed chiefly at 195 K, the samples were then warmed to various holding temperatures (for a few minutes) and, finally, measurements were made down to 4.2 K and up to 21 kOe. Surface coverage was kept relatively low. We may, therefore, have considerable confidence in the reproducibility and significance of the results.

For ethane adsorbed at 195 K and then held at 298 K (this procedure being used to minimize excessive dissociation caused by transitory heating) the saturation measurements gave a bond number (ζ)* of approxi-

* The symbols used by Martin and Imelik[6] are different from those used in this book, and certain definitions are slightly different. We use $\epsilon_{Ni}(H)$ for the change of Bohr magneton number caused by the adsorption of one mole of H atoms on one mole of nickel. Our bond number for ethane $\zeta_{Ni}(C_2H_6) = \epsilon_{Ni}(C_2H_6)/\epsilon_{Ni}(H)$. Martin and Imelik use α for the equivalent of our $\epsilon_{Ni}(H_2)$ hence for hydrogen $\alpha = 2\epsilon$, but for ethane $\alpha = \epsilon$. For bond number they use $n = \alpha/0.606$ which is larger by about 16% than our ζ. This difference is barely significant. Another difference is that Martin and Imelik use $\beta =$ the Bohr magneton $= eh/4\pi mc$. We use m_B for this quantity. We use β for the dimensionless Bohr magneton number $m(Ni)/m_B$.

mately 6.4 but rising almost vertically with holding temperature to about 12.5 at 323 K. The effect of holding temperature on average bond number over a wide range is shown in Fig. 53. (We repeat that the adsorption is carried out at 195 K, the sample is warmed to the indicated holding temperature, then the magnetic measurements are made at 4.2 K.)

The results of Martin and Imelik are in agreement with those obtained with the low-field permeameter if the transitory rise in temperature is allowed for. They also show that, at least for this system, no major difficulty arises with the permeameter results.

It will be noticed from Fig. 52 that if the metal surface is partly covered with preadsorbed hydrogen, then the fraction of chemisorbed ethane becomes negligible, although appreciable physical adsorption is still present—much of it no doubt on the silica support. This result is in agreement with the infrared results of Eischens and Pliskin.[3]

From Fig. 53 it is seen that ethane cannot be chemisorbed on nickel without dissociation. For a holding temperature below about 250 K the adsorption is physical. But, as the holding temperature is raised, dissociative chemisorption occurs in steps. Complete dissociation corresponding to a reaction

$$C_2H_6 + 12\,Ni \rightarrow 2\,Ni_3C + 6\,NiH$$

occurs at only moderately above room temperature, with the involvement of 12 nickel atoms per molecule.

Agreement of these conclusions with those of Eischens and Pliskin and of Kemball has been mentioned above. Other directly comparable studies are

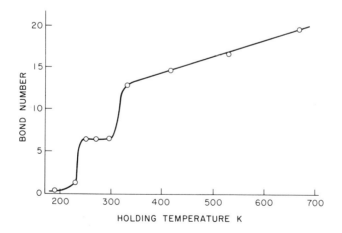

FIG. 53. Bond number as a function of holding temperature for ethane on nickel–silica (after Martin and Imelik, Ref. 6).

few in number. Freel and Galwey[7] in a kinetic study of hydrocarbon cracking reactions on nickel carbide (but containing metallic nickel) reached the conclusion that crystalline nickel carbide was not necessarily an intermediate in nickel-catalyzed cracking and hydrocracking reactions and that possibly an especially active form of carbon, obtained from dissociation of the hydrocarbon, was the source of the abundant methane formed in such reactions. Perhaps an especially active form of nickel carbide would be equally possible. The formation of methane under these conditions, has, of course, been known for many years.

Whalley et al.[8] have studied ethane on nickel by field-emission microscopy (FEM). The results are consistent with the view that H atoms and residues, including carbide, remain on the surface. In all work on this, and similar, systems it is important to remember that if surface coverage becomes appreciable, or if another reactive vapor is introduced, the surface chemistry will be drastically changed. Figure 52 shows an example of this in which preadsorbed hydrogen inhibits completely the chemisorption of ethane. But it is now possible to say that for low surface coverages of pure ethane on nickel we are beginning to have some understanding of the chemistry, if not the physics, of the process.

It may be thought that for other saturated hydrocarbons the mechanism of chemisorption would be similar to that of ethane and this, with some differences, is the case. We shall content ourselves with quoting from Martin and Imelik[6] additional data giving the approximate minimum holding temperature (with initial adsorption at 195 K) at which dissociation to hydrogen and carbide appears to be complete. Ethane is included but cyclohexane will be deferred for the next chapter.

Table IX shows that the complete cracking temperature of ethane, propane, and butane increases in that order and that methane is much

TABLE IX

CALCULATED AND OBSERVED COMPLETE CRACKING HOLDING TEMPERATURES
FOR SEVERAL PARAFFINS

| Hydrocarbon | Formula | Bond number | | Minimum holding (K) |
		Calc	Obs	
Methane	CH_4	7	6.9	573
Ethane	C_2H_6	12	12.5	323
Propane	C_3H_8	17	18.0	363
Butane	C_4H_{10}	22	18.5	373
Cyclopropane	C_3H_6	15	17.0	393

more stable than the others listed. These results will come as no surprise to any specialist in hydrocarbon chemistry.

2. The Chemisorption of Ethylene

Few problems in surface chemistry have been more hotly debated than the adsorption and hydrogenation mechanisms for ethylene. If we are still far from a complete understanding of these complex reactions we are, at least, beginning to see some agreement as to specific modes of adsorption under certain defined conditions. The history of the problem will first be surveyed briefly.

In Eley's review[9] of the catalytic hydrogenation of ethylene, he describes many different suggested modes of adsorption. Examples of these include the associative bond

$$
\begin{array}{cc}
\mathrm{H_2C} & \!\!\!-\!\!\! & \mathrm{CH_2} \\
| & & | \\
\mathrm{Ni} & & \mathrm{Ni}
\end{array}
$$

and several alternative dissociative mechanisms such as the following.

$$
\begin{array}{ccccccc}
\mathrm{HC}\!\!=\!\!\mathrm{CH_2} & \mathrm{H} & & & \mathrm{H} & \mathrm{HC}\!\!=\!\!\mathrm{CH} & \mathrm{H} \\
| & | & \text{and} & & | & | \quad | & | \\
\mathrm{Ni} & \mathrm{Ni} & & & \mathrm{Ni} & \mathrm{Ni}\ \mathrm{Ni} & \mathrm{Ni}
\end{array}
$$

Other possibilities will suggest themselves.

Most of the evidence upon which these models were based consisted of kinetic and thermodynamic studies and calculations. These, together with deuterium exchange studies, such as those described by Kemball,[10] did not succeed in establishing any one mechanism of adsorption—much less any one mechanism of hydrogenation. Most authorities in the field agreed that the presence of acetylenic residues had been proved, all agreed that self-hydrogenation involving adsorbed ethylene occurs, all agreed that adsorbed ethylene can poison nickel for the H_2–D_2 exchange reaction, and it has long been known that at higher temperatures carbiding may occur. All of these observations suggest a greater or lesser degree of dissociative adsorption, and yet the simple associative picture has an attractiveness which makes it hard to abandon.

The infrared absorption spectrum of ethylene adsorbed on nickel–silica was studied by Pliskin and Eischens.[3] On a nickel–silica sample evacuated at elevated temperature, and hence presumably bare, the intensity of the absorption bands characteristic of C–H stretching vibrations associated with saturated carbon is small compared with those bands observed when preadsorbed hydrogen is present. These results suggest that most of the ethylene is adsorbed in a manner other than associative. On a surface partially covered with preadsorbed hydrogen there is also observed a band thought to be due to a scissorlike vibration of H–C–H, and hence showing the presence of at least two hydrogen atoms on the carbon. The situation on bare nickel is further complicated by the observation that the carbon atoms appear to be saturated, even though the ratio of hydrogen to carbon atoms is quite low. This seems to suggest either polymerization, or possibly bonding to two nickel atoms by each carbon as follows:

$$
\begin{array}{ccccccc}
\text{H} & \text{HC}\!\!-\!\!\!-\!\!\!-\!\!\!-\!\!\text{CH} & \text{H} \\
| & & | \\
\text{Ni} & \text{Ni}\quad\text{Ni}\quad\text{Ni}\quad\text{Ni} & \text{Ni}
\end{array}
$$

rather than

$$
\begin{array}{cccc}
\text{H} & \text{HC}\!=\!\text{CH} & \text{H} \\
| & |\quad\ | & | \\
\text{Ni} & \text{Ni}\ \ \text{Ni} & \text{Ni}
\end{array}
$$

but the associative mechanism cannot be entirely excluded, especially for surfaces already partially covered with hydrogen. Before continuing with later infrared, and other techniques, we shall present the fairly extensive results obtained by several groups by the magnetic method.

Early results of Broeder et al.[11] for C_2H_4 admitted (apparently at room temperature) to an impregnated Ni–SiO$_2$ gave a fractional change of magnetization, by the low-field method, of the same sign and stated to be of the same order as given by H_2 on the same sample. Actually the data show a somewhat larger effect for C_2H_4. Similar studies by Selwood,[12] illustrated in Fig. 54, gave an isotherm for adsorption at 306 K with initial

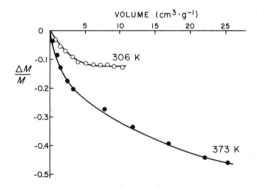

FIG. 54. Low-field magnetization–volume isotherms for ethylene on nickel–kieselguhr.

slope slightly greater than that of H_2 but sharply increased for adsorption at 373 K. These results seemed to indicate some dissociative adsorption at room temperature and extensive dissociation at moderately elevated temperature. The bond numbers (ζ) calculated from these early results at low field were 3.6 for adsorption at 306 K and 8.4 at 373 K. (Some inconclusive results for adsorption at 273 K appeared to indicate an associative mechanism, but the situation was complicated by the large amount of physical adsorption at this temperature.)

It might be thought that the maximum number of bonds which could be formed by adsorbed ethylene would be ten, representing two carbon atoms, or carbide ions, each held by three bonds, plus four Ni–H bonds, thus

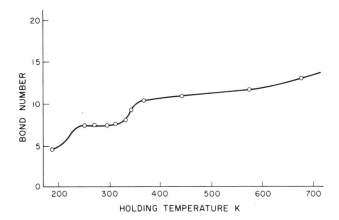

The initial slope of the magnetization–volume isotherm is actually fairly close to this value for adsorption at temperature slightly above 373 K, but still other complications may arise. These are self-hydrogenation and complete conversion of the nickel to nickel carbide. Both reactions have been known for a long time.

The magnetization changes produced by C_2H_4 on Ni have more recently been studied, under saturation conditions, by Martin and Imelik.[6] These authors also used the precaution of introducing the C_2H_4 at 195 K to minimize the effects of local heating and to make possible the observation of transitory effects. Their results are summarized in Fig. 55 which shows

FIG. 55. Bond number as a function of holding temperature for ethylene on nickel-silica (after Martin and Imelik, Ref. 6).

bond number versus holding temperature. These results, in brief, show a minimum bond number corresponding to about 4 even somewhat below 273 K, and rising in steps to 10.2 at, and above, 348 K. Our conclusions from the available magnetic data are, therefore, that associative adsorption remains uncertain and that if it occurs it is by no means the primary mode above 273 K. Complete dissociation to Ni–H and Ni_3C occurs at and above 348 K. In the room-temperature range there is magnetic evidence for a bond number of about 4, corresponding, for instance, to a mode such as

$$
\begin{array}{ccccc}
 & \overset{\displaystyle H}{|} & \overset{\displaystyle H}{|} & \\
H & C & = & C & H \\
| & | & & | & | \\
Ni & Ni & & Ni & Ni
\end{array}
$$

although there are other possibilities. It must be emphasized again that the metal particle size and the manner in which the adsorbate is admitted are all important, especially for a molecule as relatively fragile as ethylene. The heat of adsorption is sufficient to raise the temperature of a small particle many degrees and hence to cause a change in the expected degree of dissociation.

In recent years there have been many studies of ethylene chemisorption by techniques other than the magnetic. The chief results will be summarized. Morrow and Sheppard[13] observed infrared bands characteristic of the associative mode for adsorption at 195 K but this went over to a dissociative mode after 6 hr at the same temperature. (This possibility of a time-dependent change is something that has not been investigated magnetically.) For adsorption at higher temperatures the infrared and magnetic results appear to be in satisfactory agreement. Confirmatory evidence for extensive dissociation has also been found by Martin et al.[14] by low-energy electron diffraction. The same authors also reported the different adsorptive properties of various crystal faces, as did Whalley et al.[15] by field-emission studies. Photoemission spectroscopy of C_2H_4 adsorbed on Ni(III) ribbon by Demuth and Eastman[16] shows π-orbital bonding shifts characteristic, presumably, of π-associative bonding at low pressure and in the 150 K range. On raising the temperature of the pre-adsorbed species to 230 K an abrupt change was observed indicating the formation of an acetylenic species.

We shall complete this section with some further remarks about ethylene self-hydrogenation and with some other observations.

On a nickel film, evidence of ethane in the vapor phase above chemisorbed ethylene is rapidly obtained at room temperature. This occurs, presumably, by reaction of vapor-phase molecules of ethylene with dissociated hydrogen atoms from preadsorbed ethylene. The ethane is generally discovered in the vapor phase as soon as the pressure becomes appreciable. There are obviously two convenient methods for studying this reaction. In one method the ethylene is admitted in batches and the products present in the vapor phase, or capable of being desorbed, are evacuated and analyzed. In the other method, ethylene is allowed to flow over the catalyst, and the effluent gas is analyzed. This second method uses ethylene to sweep the metal surface free of self-hydrogenation products insofar as these are not themselves strongly adsorbed.

Nickel–silica preparations behave with respect to self-hydrogenation a little more sluggishly than do nickel films. This difference is possibly due to the slower desorption of ethane, some of which must be physically adsorbed on the silica. In any event, ethylene over nickel–silica at 331 K, up to a pressure of 40 $kN \cdot m^{-2}$, yields no evidence of appreciable self-hydrogenation. This is found to be true both by the evacuation method and by the method of flowing ethylene over the sample at atmospheric pressure; but if the experiment is repeated at quite moderately elevated temperature it will be found that a substantial fraction of the vapor over the catalyst is ethane. This is also found to be true if ethylene is allowed to flow over the sample in the neighborhood of 343 K—the effluent ethylene will contain ethane.

The reaction of self-hydrogenation has a peculiar effect on the quantity of ethylene which may be adsorbed on nickel. It has often been pointed out that the volume of ethylene which may be adsorbed on nickel is only one-half to one-third the volume of hydrogen which may be adsorbed on the same catalyst sample. Attempts have been made to interpret this in terms of the supposedly favorable lattice spacings found on some nickel crystal faces and not on others. It has been reported that desorption of the ethane produced by self-hydrogenation would permit the adsorption of a substantial additional volume of ethylene on nickel. At the time that work was done there was no clear-cut method for differentiating between physical and chemical adsorption, and there was some uncertainty concerning how much additional ethylene could be said to be chemisorbed under these circumstances. The process has, however, been completely confirmed.[12] It will be described in some detail.

It was stated above that there is no evidence of appreciable self-hydrogenation on nickel–silica at room temperature. But if a catalyst sample is covered with 7.3 cm^3 C_2H_4 per gram of Ni at 306 K and then heated to 373 K, it becomes possible to evacuate a considerable volume of ethane. If the sample is returned to 306 K after evacuation, it will be found that the nickel is now capable of sorbing a large additional volume of ethylene. Proof that this is chemisorption is obtained from the substantial additional loss of magnetization. We see, therefore, that the nickel is capable of taking up a volume of ethylene which even exceeds the maximum volume of hydrogen that may be adsorbed on the same sample. The requirement is that the temperature should be high enough so that a reaction may occur between ethylene and dissociated hydrogen and that the ethane should be removed.

As the adsorption temperature is raised still further, there appears abundant evidence that dissociation and other changes take place in the ethylene. At 360 K flowing ethylene causes the magnetization to fall to 58% of the initial value, and if hydrogen is then allowed to flow over the sample, it is found that the effluent gas contains a large amount of methane plus some ethane and a little higher hydrocarbon mixture. Comparable results may be obtained by evacuation and readmission of ethylene in the neighborhood of 373 K.

The final stage of dissociative adsorption is, of course, splitting off of all the hydrogen and rupture of the carbon–carbon bond. That this process occurs is shown by the following additional experiment. Ethylene was allowed to flow over the reduced nickel–silica sample at 399 K at a space velocity of about 1 s^{-1}. In 3 min the magnetization fell almost to zero, indicating almost complete conversion of the nickel to a nonmagnetic substance. If now hydrogen were permitted to flow over the sample, the hydrocarbon in the effluent was found to be almost pure methane, and the magnetization was in large part recovered. Evidence that this reaction of nickel and ethylene at moderately elevated temperature yields nickel carbide, Ni_3C, is found in the fact that heating the treated sample in vacuum at 628 K causes a substantial rise of magnetization (as measured at room temperature).

We may now summarize our views concerning the adsorption of ethylene on nickel. The widely divergent views held by many investigators are seen to be correctly attributed by Eley to the complexity of the system due to the various possibilities for self-reaction, decomposition, and polymerization. The supposedly limited ability of nickel to adsorb ethylene is due in part to the inhibition of the self-hydrogenation reaction at lower temperatures. The poisoning reaction is due primarily to dissociation and to carbiding. It is known that the poisoning effect of ethylene for the H_2–D_2 exchange reaction on nickel is much diminished if the ethylene is adsorbed below room temperature.

At lower temperatures the adsorption of ethylene may well be primarily associative. But the ethylene molecule is peculiarly sensitive to temperature and, presumably, to intrinsic activity of the catalyst. As the adsorption temperature rises there is a sharp rise in the production of ethane by self-hydrogenation. This reaction could hardly take place except by reaction of ethylene molecules with preadsorbed hydrogen formed by dissociation of the first ethylene molecules to strike the surface. That such dissociation takes place is confirmed by the progressive rise in the slope of the magnetization–volume isotherm for ethylene with rising temperature of adsorption.

As the adsorption temperature becomes still higher, it is found that the products formed by sweeping the surface with hydrogen contain an increasing proportion of methane, thus proving carbon–carbon bond rupture. Finally, at temperatures only moderately over 373 K, it is found that ethylene is capable of converting not only the surface but the whole mass of nickel into nickel carbide. This nickel carbide may readily be decomposed by hydrogen or, at somewhat higher temperature, by heat alone. This recovery of the nickel cannot, however, be achieved without some structural change in the catalyst such as growth of nickel particle size. The carbiding reaction is probably a chief offender in the poisoning action of adsorbed ethylene.

It will be noted that all of the several actions of ethylene on nickel have been proposed previously and have been the subject of investigation. The production of methane and the carbiding reaction were observed by Sabatier many years ago. If there has been a divergence of views it is because of failure to recognize that most of the proposed mechanisms possess an element of truth. The newer experimental approaches have not added many new concepts, but in some cases they have provided quantitative confirmation. The only major new idea is that there may be a quasichemisorption such as that reported by Demuth and Eastman[16] for ethylene at low temperatures. If this is correct it may throw more light on the peculiar magnetic effects of nitrogen and even of krypton (see p. 114). Some important new information on this problem has recently been obtained by Dalmon et al.[16a] By measuring the saturation magnetization on a 14% Cu–Ni alloy with low surface coverage of ethylene held in the 250–300 K region they find a bond number of 1.0. (This increases rapidly at higher holding temperature.) This work appears to be a clear indication of associative π-bonding for C_2H_4:

$$
\begin{array}{c}
\mathrm{H} \diagdown \diagup \mathrm{H} \\
\mathrm{C}=\mathrm{C} \\
\mathrm{H} \diagup \diagdown \mathrm{H} \\
\vdots \\
\mathrm{Ni}
\end{array}
$$

under these conditions.

We conclude this section with Table X showing minimum holding temperatures at which complete dissociation occurs for several olefins on nickel as reported by Martin and Imelik.[6]

TABLE X

CALCULATED AND OBSERVED COMPLETE CRACKING HOLDING TEMPERATURE
FOR SEVERAL OLEFINS ON Ni

| Hydrocarbon | Formula | Bond number | | Minimum holding (K) |
		Calc	Obs	
Ethylene	C_2H_4	10	10.2	348
Propene	C_3H_6	15	14.2	363
1-Butene	C_4H_8	20	17	398
2-Butene	C_4H_8	20	18	423
Isobutene	C_4H_8	20	18.0	403
1-Pentene	C_5H_{10}	25	18.4	398

3. Remarks on the Hydrogenation of Ethylene

If the hydrogenation mechanism for ethylene on nickel is still obscure, we may, at least, see a reason for this state of affairs. If one grants that the ethylene must first be adsorbed (a circumstance which is by no means certain), then the numerous possible modes of adsorption make it difficult to formulate any one mechanism for all conditions. We have seen how the mode of adsorption depends upon temperature, ethylene pressure, intrinsic activity of the catalyst and, we may be sure, surface coverage with hydrogen or of ethylene. In principle almost any mode of adsorption not involving polymerization or C–C bond rupture might lend itself to the hydrogenation reaction. The tortured history of this debate has recently been reviewed by Robertson.[17]

The idea put forward originally by Horiuti and Polanyi[18] was that the ethylene is associatively adsorbed and the hydrogen dissociatively adsorbed. Adsorbed C_2H_4 then reacts with adsorbed H to form adsorbed C_2H_5, which reacts with another adsorbed H to form C_2H_6 which is desorbed. A modification proposed by Twigg[19] calls for direct reaction of H_2 with adsorbed C_2H_4 to form adsorbed C_2H_5 and adsorbed H.

We shall not take space to discuss the many different mechanisms which have been suggested, but will mention one other. Jenkins and Rideal[20] suggested, in some contrast to the ideas mentioned above, that C_2H_4 from the gas-phase reacts directly with adsorbed H to form adsorbed C_2H_5 or free C_2H_6. This idea is actually a modification of Beeck's theory[21] that molecular C_2H_4 reacts with adsorbed H on a surface already partly covered with acetylenic residues derived from preadsorbed C_2H_4. Evidence for the presence of a half-hydrogenated state (adsorbed C_2H_5) is found in the infrared absorption studies of Eischens and Pliskin.[3] But several authors,[13] including Erkelens and Liefkens,[22] report that C_2H_4 on bare Ni gives the same spectrum as C_2H_4 on an H_2 covered surface.

The contribution of the magnetic method to this problem has thus far been small. It will be recalled that at room temperature on nickel–silica the ethylene molecule shows some evidence for partial dissociation. (In a typical run, the slope of the magnetization–volume isotherm is found to be greater than that for hydrogen.) Now if ethylene is preadsorbed at room temperature and then hydrogen is added over the ethylene, we obtain the isotherm shown in Fig. 56.

There are two interesting features to this isotherm. The first is that it is a straight line with a slope 2.4 times smaller than that of hydrogen alone on the same surface. (This is a good example, in contrast to the results described on p. 86, of hydrogen reacting with a preadsorbed molecule.)

The second point of interest is that the total volume of hydrogen taken up is just enough to hydrogenate all the ethylene present and to cover the surface with a monolayer of hydrogen. It is obvious that the reaction taking place is no mere replacement of a Ni–C bond by a Ni–H bond. Such a reaction would yield no change of magnetization.

The most probable explanation for these results is that the ethylene has covered only about one-third of the nickel surface which is normally accessible to hydrogen. Then when the hydrogen is added to the surface partially covered by preadsorbed ethylene, two molecules of hydrogen are adsorbed directly on to bare sites, one molecule of hydrogen is used to hydrogenate an ethylene molecule, and one molecule of hydrogen is adsorbed on the pair of sites vacated by the ethylene. This mechanism

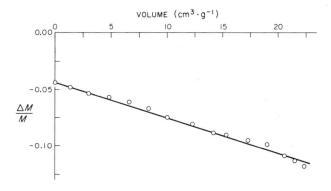

FIG. 56. Low-field magnetization isotherm for hydrogen over preadsorbed ethylene on nickel–kieselguhr at room temperature.

would lead to an isotherm slope for the hydrogen over preadsorbed ethylene just half that of hydrogen on a bare surface. However, if we recall that some dissociation by the preadsorbed ethylene has occurred and that this gives an isotherm slope a little greater than that of hydrogen alone, then the expected ratio of slopes is quite near that actually found, namely, 1/2.4 instead of $\frac{1}{2}$. It is sometimes stated that because ethylene has a higher heat of adsorption on nickel it must be more strongly adsorbed, but this cannot necessarily be true on a surface already partly covered by hydrogen.

It will be noted that the conclusion reached above is not necessarily inconsistent with the transitory existence of a half-hydrogenated state, and it is not inconsistent with the view that ethylene is normally hydrogenated directly from the vapor state. All we have shown by the magnetic method is that over preadsorbed ethylene the hydrogen is used at random, partly for hydrogenation and partly for covering bare sites. Actual hydrogenation as, for instance, carried out routinely in organic chemistry generally calls for abundant hydrogen. This must mean that the metal surface is fully covered and that any unsaturated hydrocarbon must have great difficulty finding an appropriate adsorption site. Viewed in this way the concept of direct reaction of ethylene from the vapor phase with adsorbed hydrogen atoms has considerable attractiveness. This is so particularly when we have seen how readily, at quite moderate temperatures, it is possible to recover a fully carbided nickel catalyst. In view of all this we see that this kind of magnetic experiment does not tell us much concerning the practical aspects of catalytic hydrogenation. But it would appear that the magnetic monitoring of a nickel catalyst while it is actually functioning would be an instructive experiment, and there seems no reason why this type of study should not be extended to liquid-phase reactions. It is obvious that, without exception, heterogeneous catalysis involves at least two possible surface reactants even if one is merely the product from the first.

4.   Acetylene

If attention has been lavished on the adsorption of ethylene this is less so for acetylene. The subject is reviewed (to 1961) by Bond.[23] There can be no doubt that acetylene is chemisorbed on nickel in the room-temperature region and higher. Conclusions based on reaction kinetics lead to the suggestion[24] that the adsorption mode is associative

$$HC = CH$$
$$|\quad\ \ |$$
$$Ni \quad Ni$$

analogous to the associative mode for ethylene. While not denying this possibility under certain conditions, we must not be misled into thinking that this is the only possible mechanism on a bare nickel surface over a wide range of temperature.

Infrared results by Eischens, by Little, and others are tabulated by Little.[25] There appears to be evidence for $Ni-C_2H_5$, for complete dissociation down to Ni_6C_2, and for $Ni-C_4H_9$ on a surface with H_2 added. There is no evidence for associative bonding.

An early low-field magnetic study by Broeder et al.[11] gave a result consistent with associative bonding plus a moderate amount of dissociation. More recently, Martin and Imelik[6] have made a saturation study of C_2H_2 both with and without H_2. The experimental conditions were those already described for ethane and ethylene. When C_2H_2 was adsorbed (at low coverage) on bare Ni at 195 K the bond number was 2.0, and this remained constant at holding temperature up to 253 K. This very interesting result argues strongly for an associative mode over this temperature range. The same result was observed over a somewhat limited range of surface coverage, with no evidence of self-hydrogenation. The isotherm obtained for adsorption at 195 K is shown in Fig. 57. Martin and Imelik have also extended their C_2H_2 studies over a range of holding temperature up to complete cracking at 423 K, and beyond. They have also included measurements over Ni both with preadsorbed H_2 and with H_2 over preadsorbed C_2H_2. In these cases the bond number appeared to be approximately the same, namely 2. These results are summarized in Fig. 58 with holding temperature up to 773 K. As expected, extensive, and finally complete, dissociation occurs as the temperature is raised. Holding temperatures for complete cracking of several related compounds are given in Table XI.

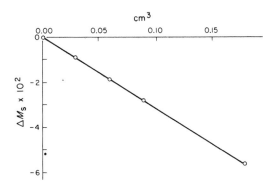

FIG. 57. Magnetization change (ΔM_s)–volume isotherm for acetylene on nickel–silica at 195 K (after Martin and Imelik, Ref. 6).

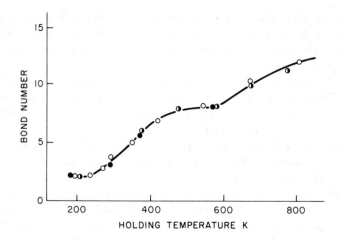

FIG. 58.   Bond number as a function of holding temperature for (O) acetylene, (●) hydrogen over acetylene, and (◖) acetylene over hydrogen, all on nickel–silica (after Martin and Imelik, Ref. 6).

Certainly one of the most interesting features of this work is the evidence of associative bonding for acetylene, and for certain related compounds, provided that the temperature does not rise above 253 K. As shown in Fig. 53. ethane shows no evidence of anything but physical adsorption under these conditions, while ethylene shows a bond number which is certainly above zero. For a clue to the reason for these differences we turn again to the recent photoemission observations of Demuth and Eastman.[16] These authors present evidence for π-bonding shifts for C_2H_2 and C_2H_4 (and C_6H_6) but none for the saturated hydrocarbons, which give no indication of bonding until, at higher temperatures, dissociation occurs.

TABLE XI

CALCULATED AND OBSERVED COMPLETE CRACKING HOLDING TEMPERATURES
FOR SEVERAL ACETYLENIC HYDROCARBONS ON Ni

| Hydrocarbon | Formula | Bond number | | Minimum holding (K) |
		Calc	Obs	
Acetylene	C_2H_2	8	7.8	423
Propyne	C_3H_4	13	12.7	423
2-Butyne	C_4H_6	18	18.0	423

The evidence for a quasichemical type of bonding is thus seen to be accumulating.

References

1. B. M. W. Trapnell, *Trans. Faraday Soc.* **52**, 1618 (1956).
2. J. R. Anderson, *Rev. Pure Appl. Chem.* **7**, 165 (1957).
3. R. P. Eischens and W. A. Pliskin, *Advan. Catal.* **10**, 1 (1958).
4. P. W. Selwood, *J. Amer. Chem. Soc.* **79**, 3346 (1957).
5. J. R. Anderson and C. Kemball, *Proc. Roy. Soc. Ser. A* **223**, 361 (1954).
6. G.-A. Martin and B. Imelik, *Surface Sci.* **42**, 157 (1974).
7. J. Freel and A. K. Galwey, *J. Catal.* **10**, 277 (1968).
8. L. Whalley, B. J. Davis, and R. L. Moss, *Trans. Faraday Soc.* **67**, 2445 (1971).
9. D. D. Eley, "Catalysis" (P. H. Emmett, ed.), Vol. III, p. 64. Van Nostrand-Reinhold, Princeton, New Jersey, 1955.
10. C. Kemball, *Proc. Chem. Soc. London*, p. 264 (1960).
11. J. J. Broeder, L. L. van Reijen, and A. R. Korswagen, *J. Chim. Phys.* **55**, 37 (1957).
12. P. W. Selwood, *J. Amer. Chem. Soc.* **83**, 2853 (1961).
13. B. A. Morrow and N. Sheppard, *Proc. Roy. Soc. Ser. A* **311**, 391 (1969).
14. G.-A. Martin, G. Dalmai-Imelik, and B. Imelik, "Adsorption-Desorption Phenomena" (F. Ricca, ed.). Academic Press, New York, 1972.
15. L. Whalley, B. J. Davis, and R. L. Moss, *Trans. Faraday Soc.* **66**, 3143 (1970).
16. J. E. Demuth and D. E. Eastman, *Phys. Rev. Lett.* **32**, 1123 (1974).
16a. J.-A. Dalmon, G.-A. Martin, and B. Imelik, *C. R. Acad. Sci. Ser. C* **279**, 1481 (1974).
17. A. J. B. Robertson, "Catalysis of Gas Reactions by Metals." Logos Press, London, 1970.
18. J. Horiuti and M. Polyani, *Trans. Faraday Soc.* **30**, 1164 (1934).
19. G. H. Twigg, *Disc. Faraday Soc.* **8**, 159 (1950).
20. G. I. Jenkins and E. K. Rideal, *J. Chem. Soc.* p. 2490 (1955).
21. O. Beeck, *Disc. Faraday Soc.* **8**, 118 (1950).
22. J. Erkelens and Th. J. Liefkens, *J. Catal.* **8**, 36 (1967).
23. G. C. Bond, "Catalysis by Metals," p. 281. Academic Press, New York, 1962.
24. J. Sheridan, *J. Chem. Soc.* p. 373 (1944).
25. L. H. Little, "Infrared Spectra of Adsorbed Species," p. 125. Academic Press, New York, 1966.

XI

Benzene and Related Compounds

1. Cyclohexane

There have been few magnetization studies on adsorbed cyclohexane. The writer[1] reported low-field isotherms over Ni–SiO$_2$ at 298 and at 423 K. At the lower temperature considerable physical adsorption was present but it was possible to conclude that some chemisorption occurred and that this was partially dissociative. At the higher temperature dissociation was extensive. The isotherms are shown in Fig. 59.

Results on cyclohexane have also been reported by Martin and Imelik.[2]

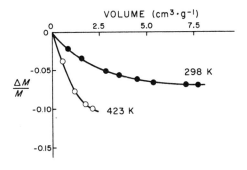

Fig. 59.   Low-field magnetization–volume isotherms for cyclohexane on nickel-kieselguhr.

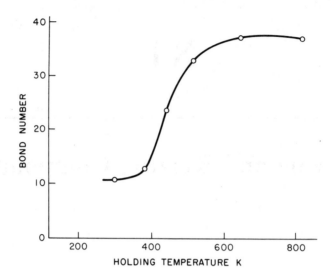

FIG. 60.   Bond number as a function of holding temperature for cyclohexane on nickel-silica (after Martin and Imelik, Ref. 2).

The adsorbate was admitted to Ni–SiO$_2$ at 195 K and at relatively low surface coverage. The sample was then warmed to various holding temperatures prior to measurement at or below 4.2 K and at high field. The results are shown in Fig. 60 in which bond number (i.e., number of Ni atoms demagnetized per molecule of adsorbate) is given as a function of holding temperature. It will be seen that dissociation is virtually complete at a holding temperature of 483 K with a bond number under these conditions of 30 corresponding to 12 Ni–H bonds and 6 Ni$_3$C bonds. As was the case for other saturated hydrocarbons studied by Martin and Imelik there was some dissociation at about 300 K. The measurements were not extended to lower temperatures at which only physical adsorption is to be expected. Within the limited range of the measurements the results are in quantitative agreement with those obtained by Galwey and Kemball[3] for deuterium exchange on a similar system.

2.   Cyclohexene

The only magnetization measurements on adsorbed cyclohexene that the writer is aware of are those of Den Besten and Selwood.[4] Figure 61 shows these data at 273 and 393 K. Hydrogen isotherms at the corresponding temperatures are also shown. At 393 K the maximum pressure over the sample was 9×10^2 N·m^{-2}. The initial slope of the isotherm obtained at

298 K was the same as that at 272 K, namely, 2.6 times that of hydrogen
under the same conditions. At 393 K the initial slope was approximately
four times that of hydrogen. These results indicate that in the temperature
range 273–298 K the average molecule of cyclohexene forms about 5.3
bonds with nickel, and that at 393 K as many as eight bonds are formed.

Before proceeding to the interpretation of these data we shall refer to the
paper by Galwey and Kemball.[5] The absorbent was a nickel–silica similar
to that used in the magnetic studies. While it is not possible to estimate the
extent of surface coverage with cyclohexene it is probable that the coverage
was comparable with that shown in Fig. 61 for the first increment or two.
The number of hydrogen atoms readily exchangeable for deuterium was
then determined in the manner to which reference has already been made
on several occasions. The results stated are that two hydrogen atoms per
molecule of cyclohexene were exchanged after adsorption at 273 K or
higher, and that four more hydrogen atoms were exchanged when the
sample was heated in excess deuterium to 393 K. The lower temperature
experiment is compaɪable with the magnetic, the 393 K run rather less so.
Results were also reported on the exchange observed for cyclohexene adsorbed at 283 K, then heated to various temperatures up to 453 K. The
number of exchangeable hydrogen atoms remained at three in this case.
There are no magnetic data strictly comparable to this last exchange result.
It has, however, been found by the writer that a hydrocarbon molecule
adsorbed at a lower temperature is somewhat more resistant to (further)
dissociation on progressive heating than is a molecule admitted to the
adsorbent at a more elevated temperature. The reason for this is doubtless
that during the adsorption process considerable excess energy is made
available during formation of the Ni–H and Ni–C bonds, and that this

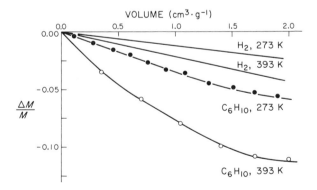

FIG. 61. Low-field magnetization–volume isotherms for hydrogen and for cyclohexene
on nickel–silica at 273 and 393 K.

excess energy may accelerate dissociative processes. The precision claimed by Galwey and Kemball is ±0.5 times the number of H atoms exchanged per molecule adsorbed. The precision of the magnetic data is believed to be somewhat better than this and perhaps the equivalent of ±0.1.

The mechanism of adsorption suggested by Galwey and Kemball for the room-temperature case is dissociative, yielding two hydrogen atoms and a C_6H_8 group. These may be adsorbed as follows:

Although precisely how the C_6H_8 radical is adsorbed is not completely established by any of these results. In any event, the concept of two hydrogen atoms being dissociated is, within probable experimental error, in agreement with the magnetic data, especially if we recall that complete fragmentation of a small fraction of the cyclohexene molecules could have a marked effect on the slope of the magnetization isotherm. The apparent total number of bonds formed per adsorbed molecule is 5.3 ± 0.5 (magnetic) and 4.0 ± 1.0 (exchange). There is no reason to believe that this mechanism of adsorption is related to the well-known disproportionation of cyclohexene over nickel–silica. The disproportionation occurs at temperatures above those at which the number of dissociated hydrogen atoms is limited to two.

Another kind of evidence offers further support for the views expressed above. While adsorbed cyclohexene may be hydrogenated at moderately elevated temperatures, it is definitely not hydrogenated at 195 K. In this experiment a pressure–volume isotherm for hydrogen is obtained at 195 K. The sample is then heated and evacuated to remove all the hydrogen, after which a measured volume of cyclohexene is admitted at room temperature. The sample is now cooled to 195 K again, and a second pressure–volume isotherm for hydrogen is obtained (Fig. 62). In a typical run it was found[4] that adsorption of 1.50 cm³ (STP) of cyclohexene vapor diminished the volume of hydrogen adsorbed (1 atm) at 195 K by 3.10 cm³, all volumes being given per gram of nickel. These results show that one molecule of cyclohexene chemisorbed at room temperature is able to deny access to the surface of two molecules of hydrogen—the hydrogen being, of course, admitted under conditions which prohibit the hydrogenation reaction. This result is in complete agreement with the views expressed above concerning the mechanism of cyclohexene adsorption. It was earlier pointed out (see p. 86) that the presence of preadsorbed cyclohexene (at 195 K) does not alter the slope of the magnetization–volume isotherm for hydrogen.

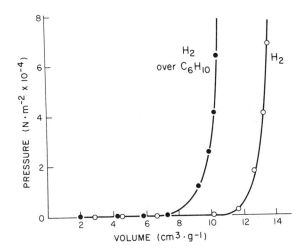

FIG. 62. Pressure–volume isotherms for hydrogen and for hydrogen over cyclohexene on nickel–kieselguhr at 195 K.

3. The Chemisorption of Benzene

The interaction of benzene, and of benzene plus hydrogen, on nickel have been studied almost as exhaustively as has ethylene. We shall first present the available magnetization evidence for benzene alone and then attempt to reconcile our conclusions with those from the wealth of other data.

Low-field isotherms obtained by Silvent and Selwood[6] for C_6H_6 on Ni–SiO$_2$ over a range of temperature are shown in Fig. 63. At room temperature, physical adsorption is large as soon as the pressure becomes appreciable. This is responsible at least in part for the slopes diminishing with increasing coverage at the lower temperature. It will be noted that the total volume, measured as vapor corrected to standard conditions, is much smaller for benzene than for hydrogen. The first increments of benzene show strong thermal transients like those shown by hydrogen (see p. 90) and other adsorbates.

In spite of these several complications it is clear that benzene enters into electronic interaction with nickel over the whole temperature range represented in Fig. 63. It is also clear that the temperature of adsorption has a marked effect on the number of bonds formed per molecule of adsorbed benzene. The results may be interpreted as being suggestive of associative adsorption at lower temperatures and dissociative at higher. Six bonds formed by benzene suggest that the molecule may lie flat; more than six

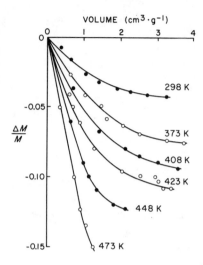

FIG. 63. Low-field magnetization–volume isotherms for benzene on nickel–kieselguhr.

bonds suggest not only dissociation of hydrogen but some degree of carbon–carbon bond rupture.

Some evidence that these views may be correct are obtained from other experiments conducted parallel to the magnetic studies. The first is the rather obvious one of attempting to hydrogenate the adsorbed molecules or molecular fragments. It is found that if hydrogen is allowed to flow over preadsorbed benzene at 373 K, it is possible to recover, by freezing, most of the original adsorbate in the form of cyclohexane. But if the adsorption step is conducted at 473 K, the effluent vapor contains an appreciable fraction of lower molecular weight hydrocarbons, although a substantial amount of cyclohexane still appears. This result suggests that the very large number of bonds indicated by the magnetic data for the higher adsorption temperatures is made up of hydrogen–nickel bonds plus carbon–nickel bonds, but with the carbon still mostly present in the form of six-membered rings.

Additional evidence supporting these views is obtained from another experiment, which is to find the volume of hydrogen denied access to the nickel (at 195 K) by the presence of a measured volume of preadsorbed benzene, as already described for cyclohexene (see p. 140). The result of this experiment is that at a benzene adsorption temperature of 298 K one molecule of benzene denies access to the nickel of 2.6 molecules of hydrogen. This number rises with increasing benzene adsorption temperature until at 473 K it is 8.6.

We turn now to the saturation magnetization study of Martin and

Imelik[2] carried out under carefully controlled conditions. Figure 64 shows the bond number obtained as a function of holding temperature over a wide range. Below 273 K chemisorption is negligible, but in the room-temperature region the bond number rises rapidly up to about 8, after which it continues to rise, reaching 25.0 (corresponding to complete dissociation to H and C) at about 473 K. The high- and low-field methods are, therefore, in reasonably close agreement concerning the experimental facts. (Some additional data of Martin and Imelik on the $C_6H_6 + H_2$ system will be deferred until the following section.)

The chemisorption of benzene has recently been reviewed by Moyes and Wells.[7] The major conclusions, based on a wealth of experimental data, are that associative chemisorption as a π-complex occurs, that adsorption (in the absence of hydrogen or other contaminants) leads to fission of at least two carbon–hydrogen bonds, and that further dissociation occurs as the temperature is raised. With these conclusions the magnetic data are in qualitative agreement although the exact temperatures and other conditions may vary from case to case. Moyes and Wells (Ref. 7, p. 130) question whether the magnetic data may actually be interpreted as flat and six-bonded for adsorption at the lower temperatures. This doubt appears to be based on a failure to realize that loss of some magnetization in superparamagnetic particles does not necessarily involve donation of electrons to the adsorbent. We also mention the evidence obtained by Demuth and Eastman[8] from photoemission spectroscopy that benzene (like C_2H_2 and C_2H_4), adsorbed at low temperatures, exhibits the π-orbital bonding shifts consistent with an associative mode.

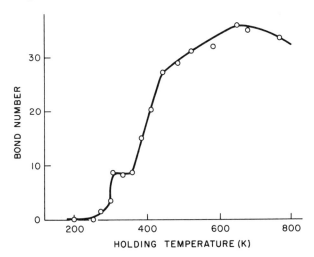

FIG. 64. Bond number as a function of holding temperature for benzene on nickel-silica (after Martin and Imelik, Ref. 2).

Our final conclusions with respect to benzene on nickel are in general agreement with those listed by Wells and Moyes, but an area of ambiguity remains. Garnett[9] shows that π-complex adsorption may be either associative or dissociative. The ring lies flat

Associative	Dissociative
π-complex	π-complex

in both cases—"edge-on" adsorption being prevented by orbital symmetry and by steric hindrance, except that ring rotation may be necessary for substitution reactions. The associative complex would, presumably, hinder the nonreactive adsorption of about 3 molecules of H_2, as actually found. The dissociative complex would hinder the adsorption of 3.5 to 4 molecules of H_2, depending on the total surface coverage. The most carefully done magnetization measurements indicate about 8 Ni atoms demagnetized per molecule of C_6H_6 adsorbed in the room-temperature region. Martin and Imelik[10] interpret this as being due to "edge-on" dissociative adsorption

which would give a bond number of 8. But this mode does not agree with the Demuth and Eastman results (under quite different experimental conditions). It seems reasonable that Garnett's two flat modes indicated above would yield magnetic bond numbers of 6 for the associative and 7 for the dissociative. We cannot say which one of the alternatives seems the more probable.

In conclusion we refer again to the possibility that certain molecules, of which ethylene and benzene are only two, sometimes exhibit a kind of quasichemisorption that seems to lie between true physical and true chemisorption. This affects the magnetization and certain other properties but does not affect all the properties normally considered sensitive to chemisorptive bonding. If this occurs for molecular nitrogen (see p. 114) it certainly merits further investigation.

4. The Interaction of Benzene and Hydrogen

The conclusions reached by Moyes and Wells[7] in their comprehensive review of the subject may be summarized as follows. Benzene in competition with molecular hydrogen for a catalyst surface undergoes hydrogenation, but the stepwise deuterium exchange suggests reversible formation of adsorbed C_6X_7, possibly by a modified Rideal–Eley mechanism. This exchange appears to be structure sensitive. Aromaticity is retained during exchange but the number of delocalized π electrons is temporarily changed from 6 to 5. The hydrogenation does not require a special state of chemisorption of the benzene. The writer has not encountered any evidence showing that benzene cannot be hydrogenated unless it is first chemisorbed.

The number of magnetization studies on this subject is limited. Low-field measurements[1] show that if a nickel surface already partly covered with benzene (admitted at room temperature) is then treated with hydrogen it will be found that the hydrogen, as shown in Fig. 65, is preferentially chemisorbed rather than being used for hydrogenation. This is proved because the hydrogen isotherm slope is the same as if no benzene had been present.

Now if more hydrogen is added it will be found (Fig. 65) that the isotherm becomes more nearly horizontal, showing that no net change of bond number is occurring. This is consistent with the view that hydrogen is now being used for hydrogenation, that nickel–carbon bonds are being broken, and that the vacant nickel sites are being replaced by nickel–hydrogen bonds. The total volume of hydrogen taken up in this process is just that required to hydrogenate all the benzene, plus enough to cover the nickel. The implication is then that while benzene may readily be hydrogenated at room temperature, yet the adsorbed benzene cannot be hydrogenated directly without prior desorption. When the bond strengths have been weakened to an appropriate degree by increasing surface coverage, then, and then only, can the benzene be desorbed. As soon as the benzene is desorbed from the nickel it is free to pick up six adsorbed hydrogen atoms.

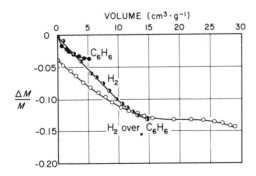

FIG. 65. Low-field magnetization–volume isotherms for hydrogen, for benzene and for hydrogen over preadsorbed benzene on nickel–kieselguhr, all at room temperature.

The strong van der Waals adsorption probably prevents much emergence of cyclohexane in the gas phase. If preadsorbed benzene cannot be hydrogenated at 195 K, it is probably because the hydrogen is unable to displace the benzene at this temperature.

Further evidence is obtained by covering the nickel surface, at room temperature, with a partial layer of hydrogen, then adding benzene as shown in Fig. 66. Under these circumstances the benzene causes a negligible change of magnetization. Yet, if any appreciable quantity of preadsorbed hydrogen were being used for hydrogenation, then the subsequent addition of more hydrogen would cause a substantial loss of magnetization. But as shown in Fig. 66, this loss does not occur. We must, therefore, conclude that in the presence of somewhat tightly bound hydrogen the benzene molecule can neither take hydrogen away from the nickel nor can it find an appropriate site for its own chemisorption. But when additional hydrogen is added, the pressure–volume relationship shows that quantitative hydrogenation readily occurs.

These several considerations make it possible to demonstrate the participation of the catalyst during an actual catalytic process. A nickel–silica sample reduced and evacuated as usual is covered to about 1 atm with hydrogen, then sealed off. A few drops of benzene previously introduced and kept frozen adjacent to the catalyst is then allowed to melt. As the benzene diffuses to the nickel surface, some hydrogenation occurs. This causes diminution of the hydrogen pressure which, in turn, leaves some nickel sites bare. The attendant rise of magnetization is readily observed as hydrogenation proceeds (Fig. 67).

The only other magnetization study of the $C_6H_6 + H_2$ system on Ni

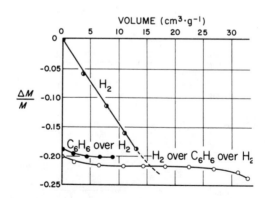

FIG. 66. Low-field magnetization–volume isotherms for hydrogen, for benzene over preadsorbed hydrogen, and for hydrogen over benzene over preadsorbed hydrogen on nickel–kieselguhr, all at room temperature.

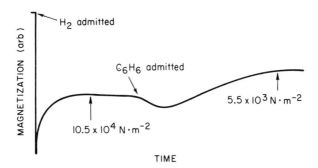

FIG. 67. Low-field magnetization changes occurring in nickel–kieselguhr at room temperature as benzene is admitted to a sample covered with preadsorbed hydrogen.

appears to be that of Martin and Imelik.[2] Their saturation measurements, with C_6H_6 admitted at 195 K as described above, show that if H_2 is admitted over preadsorbed C_6H_6 the slope of the isotherm is almost exactly the same as that for H_2 on a bare surface. But if the C_6H_6 is admitted at room temperature to a surface already covered with H_2 the isotherm is complicated starting at an apparent bond number of about 3.5 at low coverages and falling to about 0.8 at higher coverage. A summary of these results is shown in Fig. 68. These results are in reasonably satisfactory agreement

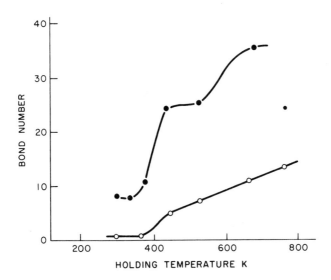

FIG. 68. Approximate bond number as a function of holding temperature for (●) benzene over preadsorbed hydrogen and (○) hydrogen over preadsorbed benzene on nickel-silica, at low H_2 coverage (after Martin and Imelik, Ref. 2).

with those obtained by the low-field method. The conclusions reached by Martin and Imelik are that when the hydrocarbon is adsorbed first there is no interaction with hydrogen provided, of course, that the temperature is kept low. But if hydrogen is first adsorbed it inhibits dissociation of the benzene but is available for interaction leading to hydrogenation. These results are consistent with the view that during hydrogenation most of the catalyst surface is covered with hydrogen and benzene reacts from the gas phase. This conclusion does not, of course, mean that the actual reaction may not involve a transient chemisorbed benzene molecule, but rather that, if this occurs, the concentration is relatively small.

References

1. P. W. Selwood, *J. Amer. Chem. Soc.* **79,** 4637 (1957).
2. G.-A. Martin and B. Imelik, *Surface Sci.* **42,** 157 (1974).
3. A. K. Galwey and C. Kemball, *Trans. Faraday Soc.* **55,** 1959 (1959).
4. I. E. Den Besten and P. W. Selwood, *J. Catal.* **1,** 93 (1962).
5. A. K. Galwey and C. Kemball, *Actes Congr. Int. Catalyse, 2e, Paris 1960* **2,** 1063 (1961).
6. J. A. Silvent and P. W. Selwood, *J. Amer. Chem. Soc.* **83,** 1033 (1961).
7. R. B. Moyes and P. B. Wells, *Advan. Catal.* **23,** 121 (1973).
8. J. E. Demuth and D. E. Eastman, *Phys. Rev. Lett.* **32,** 1123 (1974).
9. J. L. Garnett, *Catal. Rev.* **5,** 229 (1972).
10. G.-A. Martin and B. Imelik, *J. Chim. Phys.* **68,** 1550 (1971).

XII

Other Systems

1. Palladium and Platinum

Palladium has certain properties much like those of nickel except that the electrons contributing to its permanent magnetic moment are 4d rather than 3d. The adsorptive properties have been studied exhaustively and, in many respects, they resemble those of nickel. The Bohr magneton number β is about 0.6, and the atoms do not have cooperative interaction leading to ferromagnetism, and to superparamagnetism in small particles. Certainly over a wide temperature range Pd is paramagnetic. It might, therefore, be thought that a study of the magnetization effects caused by adsorbed molecules on palladium would be fruitful in helping us to understand the corresponding nickel systems.

But palladium has its own full share of complexities. The magnetic moment is the moment derived from susceptibility measurements on the paramagnetic solid and thus must be compared with the much higher moment of nickel above the Curie point (see p. 12). Furthermore, the magnetic properties of palladium become increasingly complicated at lower temperatures, and the susceptibility is extraordinarily dependent on traces of impurities such as iron. And finally, adsorbed hydrogen under most conditions passes across the surface and forms at least two solid phases becoming, to a degree, like the case of nickel subject to cathodic hydrogen (see p. 99).

In spite of these problems there have been some studies on adsorbed molecules on palladium, and these will be described. The experimental method is generally some adaptation of the Faraday method. The gradient field coil system described by Lewis (see p. 40) would appear to be especially appropriate for such studies. In the absence of ferromagnetic interaction, as in nickel, palladium has a magnetization, in fields of a few kOe, three or four orders smaller than a similar supported nickel sample. This makes it necessary to consider diamagnetic corrections for most adsorbates. The susceptibility κ/ρ of palladium is less than one order larger (arithmetically) than that of benzene.

There have been many magnetic studies on the system formed by hydrogen dissolving in palladium. In brief, the hydrogen appears to be dissociated to atoms or ions, more than one phase is formed, and the susceptibility as measured at room temperature becomes zero at a ratio of about $PdH_{0.64}$. The chief problem in a study of adsorbed H_2, rather than of absorbed H_2, is to prevent the latter from obscuring the results on the former. Another problem is that there is conflicting evidence concerning the susceptibility of finely divided, supported palladium. Trzebiatowski et al.[1] reported that Pd supported on alumina gel had a lower than normal susceptibility. Reyerson and Solbakken[2] reported a higher than normal susceptibility for silica-supported palladium. The problem of differentiating between adsorbed and absorbed may be solved, in part, by the method proposed by Aben.[3] Examination of Pd–H_2 isotherms obtained at various temperatures shows that the quantity of H_2 absorbed is minimal at, for example, 343 K and under 10^2 N·m^{-2}. Under these conditions the quantity of absorbed H_2 is no more than about 2×10^{-3} H atom per Pd atom. (This method has been successfully used by Benson et al.[4] in connection with the hydrogen–oxygen titration method for obtaining palladium surface area.) Aben's paper, although including no magnetic data, does give the useful result that under the conditions recommended about one hydrogen atom is adsorbed per surface palladium atom. The same result is *assumed* by Benson et al. and others, who have discussed the matter.

Reyerson and Solbakken have attempted to resolve this problem by measuring the rate of sorption. Adsorption is virtually instantaneous, absorption slower. This method has an advantage in that the experimental conditions are not so limited. Because of these and other complexities it is not easy to estimate the change in moment of the Pd as a function of H_2 adsorption, as was done for Ni. Figure 69 shows a part of the data presented by Reyerson and Solbakken for a silica-supported sample containing 11.36% Pd, and at 273 K. The authors did not attempt to calculate the equivalent of ϵ for this system, and we can only make a rough estimate from the published data. Recalling from Eq. (1.6) that β is proportional to $\kappa^{1/2}$ we note that the fractional change of susceptibility is 0.24 at an atom ratio

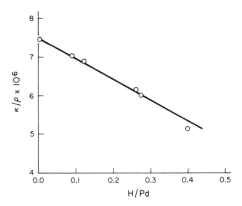

FIG. 69. Magnetic susceptibility (κ/ρ)–volume isotherm for hydrogen on palladium at 273 K (after Reyerson and Solbakken, Ref. 2).

of H:Pd = 0.4. Hence,

$$\epsilon_{Pd}(H) = (0.24)^{1/2}\beta(Pd)/0.4 = 0.7$$

Then we may say that, within at most about ±0.3, ϵ for Pd is the same as ϵ for Ni.

The magnetic susceptibility of the nitric oxide–palladium system has been studied by Solbakken and Reyerson[5] and by Zuehlke et al.[6] Here the situation is complicated by the paramagnetism of the adsorbate, but absorption is not a problem. The results show clearly that in the 77 K range the susceptibility rises with increasing surface coverage, reflecting the physical adsorption of a paramagnetic layer with, possibly, some dimerization. At higher adsorption temperatures increasing chemisorption occurs, and this becomes exclusively chemisorption, or very nearly so, until monolayer coverage is exceeded. The susceptibility falls in this case. Figure 70 shows examples of two isotherms illustrating these effects, out of the many presented by the authors. Again, it is difficult to estimate the change of moment caused by one chemisorbed nitric oxide molecule, but from the data given we believe $\epsilon_{Pd}(NO)$ to be approximately 1.2. (The authors state that about 1.5 electron is transferred to the d band of the metal but we are unable to follow the reasoning.) It may be noted that there is considerable infrared evidence for linear bonding for NO on various metals.[7]

The work of Dilke et al.,[8] to which reference has already been made, showed that adsorbed dimethyl sulfide on palladium powder, at 293 K, lowered the susceptibility (after correction for the diamagnetism of the adsorbate). Within about ±50% these preliminary, but very interesting results showed that the magnetic moment of each surface atom of pal-

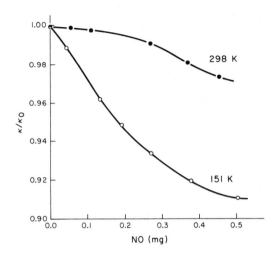

F<small>IG</small>. 70. Relative change of susceptibility with adsorbed nitric oxide on palladium (after Zuehlke *et al.*, Ref. 6).

ladium was reduced to zero (when coverage was complete) and that this apparently occurred through the donation of one electron from each molecule of the sulfide. Be that as it may, we can make a rough estimate of $\epsilon_{Pd}(C_2H_6S)$, and this is about 0.6.

We may now compare ϵ, at least for hydrogen and for dimethyl sulfide on palladium and on nickel, and for the case of hydrogen we may compare the same quantities for absorption rather than for adsorption. This is done in Table XII. (There are no data for NO on Ni.)

The data available for constructing Table XII are so limited that few definite conclusions may be drawn. However, there is no question that Table XII suggests avenues for further study. With one exception the ϵ values are astonishingly alike, and that one, dimethyl sulfide, could be attributed to differences in intrinsic activity (related to particle size and

T<small>ABLE</small> XII

ϵ F<small>OR</small> H<small>YDROGEN</small> <small>AND FOR</small> D<small>IMETHYL</small> S<small>ULFIDE</small>

	Absorbed	Adsorbed
$\epsilon_{Pd}(H)$	~ 0.6	~ 0.7
$\epsilon_{Pd}(C_2H_6S)$	—	~ 0.6
$\epsilon_{Ni}(H)$	0.6	~ 0.7
$\epsilon_{Ni}(C_2H_6S)$	—	~ 1.4

geometry of the adsorbent) or even to deficiencies in the permeameter method used for the nickel. It appears that not only do adsorbed and absorbed hydrogen produce the same ϵ for nickel, but also for palladium. But palladium does not form superparamagnetic particles. It is not safe, in this area, to place much confidence in reasoning by analogy, but it certainly is suggested that the adsorption and absorption mechanisms for hydrogen do not differ greatly for the two metals. If this is actually the case, then we may perhaps regard any loss of cooperative exchange interaction by the nickel as being incidental to filling of the d band, either locally or collectively.

From the above results it is also to be noted that the use of palladium as an adsorbent does not offer any easier path to understanding chemisorption. However, further studies of the kind described should answer more than a few questions. One of these is: What happens to the hydrogen from a partly dissociated adsorbate molecule such as ethylene? Does it stay on the surface of palladium as it does on nickel, or does it penetrate the bulk?

For platinum $\kappa/\rho \simeq 1.0 \times 10^{-6}$ cm$^3\cdot$g^{-1} at room temperature, and this low value makes even more difficult the problems of purity and the high precision necessary to achieve meaningful results. The susceptibility decreases slowly with rising temperature. Gray and McCain[9] have used the Gouy method to measure the susceptibility of platinum in the forms of foil, gauze, and powder in vacuum and oxygen and hydrogen at various pressures and temperatures. The purpose of the work, an important one, was to gain information about the nature of the catalytic activation process when platinum is treated at elevated temperature in a hydrogen–oxygen mixture. The results and conclusions will be summarized.

Oxygen, in general, decreases the susceptibility of platinum. (This effect should not be confused with the *apparent* decrease in susceptibility of any substance measured by a ponderomotive method in a paramagnetic atmosphere.) The susceptibility is restored to its original value if the sample is heated in hydrogen. For a sample evacuated at high temperature it will be found that the admission of hydrogen has no effect on the susceptibility. Various complications ensue if the details of pretreatment are altered. Some of the experimental results are shown in Fig. 71.

It is clear from these results that the behavior of platinum with respect to adsorbed hydrogen is quite different from that on nickel or palladium. The effect of oxygen appears to be more nearly like that on nickel, but with the surprising difference that one oxygen atom appears able to affect the magnetization of 10–50 atoms of platinum. But in spite of these several complications it seems quite probable that activated platinum may owe its activity to some form of metal–oxygen interaction on the surface.

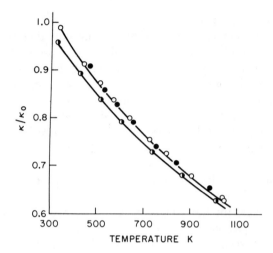

Fig. 71. Relative change of susceptibility with temperature for (○) bare platinum, (●) hydrogen on platinum, and (◑) oxygen on platinum (after Gray and McCain, Ref. 9).

2. Oxides

There are many oxides that are ferromagnetic, or ferrimagnetic, and several of these have been shown to exhibit superparamagnetism in small particles. It will be recalled that Néel's development of the theory of superparamagnetism was based on such substances, as found in nature. Provided that the particle sizes are small enough these substances may be expected to show changes in magnetic moment if chemisorbed molecules are present on the surface, and this appears to be the case. But most observations of such effects have been made incidentally to other work. Examples will be mentioned.

Many years ago Elmore[10] noted that suspended colloidal γ-Fe_2O_3 appeared to have a subnormal moment. More recently Kaiser and Miskolczy,[11] in a study of Fe_3O_4 suspended in hydrocarbon oil and stabilized with oleic acid showed a low moment which they attributed to a nonmagnetic iron oleate surface complex. Somewhat similar effects on various ferrites have been observed by Chirulescu and Segal[12] with adsorption from the gas phase. Berkowitz and Lahut[13] have also studied a large number of ferrites powdered by prolonged milling with steel balls. (This had the effect of introducing some iron for which corrections had to be made.) The results show again that surface-active molecules result in a lowered magnetization.

The above results have merely a peripheral bearing on our main problem, but they do suggest that further chemisorption work on appropriate oxides is warranted and also that liquid-phase adsorption might prove useful. Specific surfaces adequate for this purpose may be obtained by the method of attrition grinding.[14] Among the oxides that might prove to be especially interesting is chromium dioxide. With the aid of a high sensitivity permeameter, Illgen and Scheve[14a] observed an effect of adsorbed oxygen on the magnetization of nickel ferrite, $NiFe_2O_4$, of surface 51 $m^2 \cdot g^{-1}$.

3. Raney Nickel

Raney nickel is formed by treatment of nickel–aluminum alloy with a solution of sodium hydroxide. The convenience and usefulness of the resulting catalyst are more than matched by its structural complexity. A catalyst formed in this way is often referred to as a "skeletal" metal. Some authors refer to Raney nickel as a "sponge" catalyst, but the term "sponge" is generally used for a catalyst formed by coagulation of colloidal metal particles.

The elementary composition of the most widely used preparation of Raney nickel is nickel, aluminum, oxygen, and hydrogen. Substances identifiable by x-ray diffraction and other methods include, Ni, Ni_2Al_3, $NiAl_3$, certain eutectics, various hydrates of Al_2O_3, water, and hydrogen in some form not fully understood. There have been many studies on the structural composition of Raney nickel. The papers mentioned below contain references to earlier work. It is virtually certain that Raney nickel is not homogeneous. A notable property of this catalyst is that when heated it releases hydrogen in large quantities amounting, for instance, to 100 $cm^3 \cdot g^{-1}$ (STP) a part of which is reversible. The information hoped for from magnetic measurements consists primarily of (1) particle size determination, (2) composition of the nickel-rich fraction with respect to aluminum and to adsorbed hydrogen, and (3) the source of liberated hydrogen.

The grain of what is, presumably, the active metal in Raney nickel is not pure nickel. This complicates any attempt at granulometry. The problem has been attacked by Fouilloux et al.[15] and by MacNab and Anderson.[16] The former group attempted to simplify the problem by leaching all but ~3% of the Al from the sample. Saturation magnetizations were obtained on samples from which all, or nearly all, the available hydrogen had been removed. The Langevin high-field (LHF) method then yielded average particle diameters, and these were compared with results obtained by other methods. The average results, without regard to distribution, were approximately: (LHF) 10 nm, x-ray line broadening 5, small angle scattering 6, electron microscopy pore diameter 7 nm. These data give, at least, some idea of the particle diameters. Nitrogen adsorption by the BET method gave surface areas of the order of 70 mg$^2 \cdot$ g^{-1} and a, most abundant, pore diameter of about 3.0 nm.

The saturation magnetization measurements of MacNab and Anderson[16] were performed on a variety of samples of varying preparation and pretreatment. Taking into consideration the probability that residual hydrogen may lower the magnetization of the metal, and also that the metal may contain nonmagnetic matter, these authors obtain (LHF) particle diameters in the 8.0–10.0 nm range, to be compared with BET areas of 60–75 m$^2 \cdot$ g^{-1} and x-ray linewidth diameters of 4.0–10.0 nm. The general range of metal particle diameters in Raney nickel is, therefore, established about as well as could be expected for a system of this kind.

Our second problem here is to find what may be learned about the composition of the metal which is, presumably, the site of catalytic activity. If hydrogen is chemisorbed on the nickel there will certainly be some lowering of M_s. Fortunately, this possibility has been carefully investigated. Kokes and Emmett[17] found that the magnetization rose as the hydrogen was removed by progressive heating up to 773 K. (A strange feature is that the desorption is exothermic.) The change in M was about $+100\%$. This effect was reversible. Similar, more extensive, studies by Fouilloux et al.[15] are illustrated in Fig. 72 which shows M_s(arb) versus volume of H$_2$ desorbed per gram of catalyst, as measured at 4.2 K. At 300 K the isotherm is similar but, of course, with lower M. An indication of the change on reversible sorption is also given in Fig. 72. The amount of H$_2$ that can be

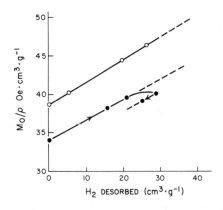

Fig. 72.   Magnetization (M_0/ρ) during desorption of hydrogen from Raney nickel (○, after McNab and Anderson, Ref. 16; ●, after Fouilloux et al., Ref. 15).

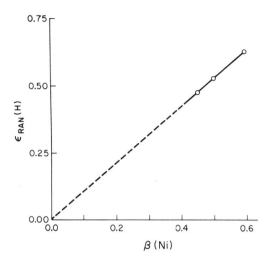

FIG. 73. Change of moment on desorption of hydrogen (ϵ) from Raney nickel as a function of the Bohr magneton number (β) of the nickel (after Martin and Fouilloux, Ref. 19).

resorbed is not large. The same authors show that the change in M_s on desorption may be interpreted as equivalent to $\epsilon_{RAN}(H) = -0.6$ which is, of course, the same as for $\epsilon_{Ni}(H)$ where the Ni is supported (or where the H is absorbed), as suggested by Kokes and Emmett. However, Fouilloux et al. find that M_s at 4.2 K is only 73% of that for pure massive nickel, and they attribute this to the presence of about 3.5% aluminum dissolved in the nickel. This may well be the case but it raises a question concerning the meaning of ϵ. The effect of dissolved copper on $\epsilon_{Ni-Cu}(H)$ has already been described (see p. 62). However, the aluminum concentration in Raney nickel was low compared with the copper concentration in the Ni–Cu alloys studied in this way.

Figure 72 also shows results obtained by MacNab and Anderson[18] on a similar, but not identical, sample. Here, too, $\epsilon_{RAN}(H)$ was nearly the same, namely, -0.63, but here again we are faced with uncertainties regarding the effect of dissolved aluminum. Some better understanding of, at least, the experimental facts, if not the interpretation, may be gained from the work of Martin and Fouilloux[19] (Fig. 73) who have plotted the saturation moment of the metal versus $\epsilon_{RAN}(H)$. The results (given in the original $\alpha = 2\epsilon$), are not extensive but they appear to show that ϵ varies linearly with β(Ni) and extrapolates to zero at β(Ni) = 0. Taken, together with the concurrent measurements on remanence and Curie points published by both groups of authors, we may interpret the data given in Figs. 72 and 73, as follows. The metal in Raney nickel is an alloy. In contrast to nickel–silica which is an array of metal particles suspended on silica, the Raney nickel may (somewhat facetiously) be thought of as an array of voids suspended on metal. The residual alumina appears to be incidental although we cannot say what, if any, catalytic influence it may have. There appear to be three choices for the source of the hydrogen liberated when the sample is heated. It may be generated from reaction of water with Ni or Al or both, it may be absorbed or it may be chemisorbed on the surface. The conclusion reached by Martin and Fouilloux[19] is that the hydrogen is present as such and mainly chemisorbed on the metal. The conclusions of MacNab and Anderson[18] are that there is too much hydrogen to be chemisorbed solely and that much of it must be absorbed or interstitial and that a substantial part may be generated by oxidation of the residual aluminum by water. It will be noted from Fig. 72 that just because the desorption isotherm is a straight or nearly straight line, does not prove that all the hydrogen comes from the same source. These latter suggestions seem more consistent with the fact that the desorption is exothermic. These later views are discussed by Nicolau and Anderson.[20]

4. Reflections

Within its narrow range of applicability the measurement of saturation magnetization is one of the most useful of all experimental methods for learning what happens to the adsorbent during chemisorption on a practical catalyst. It gives directly the number of atoms affected to the degree that they may no longer participate normally in the cooperative properties of the catalyst. By inference this reveals the extent of bonding, with or without dissociation, suffered by the adsorbate molecule. Taken in conjunction with the less reliable, but more flexible, low-field method these measurements may be extended over the whole range of experimental conditions commonly encountered in heterogeneous catalysis.

For systems to which the method may be applied no theoretical discussion can afford to ignore the magnetic results. Taken together with the wealth of results coming from reaction studies and new experimental techniques it seems not improbable that the long-sought unifying theory of heterogeneous catalysis may soon be at hand. These concluding remarks are merely to point out, again, areas where uncertainty still reigns, and to mark some possibilities for future developments.

Instrumentation for magnetic measurements on catalytically active solids has reached a satisfactory, although perhaps not fully exploited, level. In the field of granulometry the ac modification of the Néel relaxation method, as described by Martin, appears to have promise. For saturation measurements the adaptation, described by Lewis, of the field gradient magnetometer seems highly appropriate although perhaps not for fields in excess of about 20 kOe. For chemisorption studies the techniques of gas handling, as refined by Martin and his associates at the CNRS laboratories in Villeneuve, have much to recommend them. It is probably correct to say that for all adsorbates except hydrogen it is essential to admit the vapor at a temperature well below the lowest known, or suspected, dissociation temperature. This must be done with full regard for the temperature rise caused by the adsorption process, even though it may be only physical. And if, for instance, the vapor pressure of benzene at 195 K is below 10^{-2} $N \cdot m^{-2}$ so much the better. This will inhibit possible reaction of gas-phase molecules with those already adsorbed. Furthermore, the time elapsed after adsorption must be considered. This is an area that has, thus far, received little attention.

The low-field permeameter method makes it possible (for some systems) to gain information concerning the state of a catalyst while it is functioning, although there have been very few such applications. It would, for instance, be possible to measure the magnetization of nickel–silica while it is catalyzing the hydrogenation of an olefin. This obviously raises some problems because the nickel acts as its own thermometer in such cases, and the magnetization is also changed by the concentration and kind of adsorbed molecules. Solutions to this problem do not seem to be impossible.

The most interesting and useful result that emerges directly from these measurements is the number ϵ. It is useful because it permits us to find the bond number ζ for any adsorbate molecule. But it must be admitted that we have as yet a far from clear idea of the theoretical significance of ϵ. Such an understanding is not likely to come until we gain a better theoretical basis for the magnetic properties of transition metals in general and, especially, of the situation at the surface. Such a development is much to be desired. But it must be pointed out that the practical development of heterogeneous catalysis—a development that is not only all-pervasive but that can only be described as magnificent—has been made by knowledgeable scientists and engineers working with an abundance of facts but with little theoretical assistance.

The ferromagnetic, or ferrimagnetic, oxides as adsorbents and the possibility that there is a kind of quasichemisorption showing some, but not all, of the usual criteria of chemisorption are among the many areas appropriate for the investigation by the magnetic method. The study of the metals cobalt and iron is in its infancy and should be rewarding, but for the nonferromagnetic metals such as palladium and platinum it appears that the magnetic methods described in this book are less helpful than, say, Auger spectroscopy. Nevertheless, it would be interesting to learn more about systems such as palladium–dimethyl sulfide investigated by Dilke and Eley many years ago. A step in this direction has actually been taken by Saito and Fuma[21] who compared the susceptibility change of supported palladium on exposure to hydrogen before and after the metal had been covered with benzene and with α-methylstyrene.

Finally, it is not inappropriate to repeat the plea that investigators not only read the earlier literature but exercise scrupulous care in describing the systems with which they work. It is not much exaggeration to say with respect to ethylene on nickel (and to other systems) that if two experienced workers reach the same conclusions it merely means that their experimental conditions were more nearly identical than either had a right to expect. If they reach different conclusions then the chances are that both are correct.

References

1. W. Trzebiatowski, H. Kubicka, and A. Skiva, *Rocz. Chem.* **31**, 497 (1957).
2. L. H. Reyerson and A. Solbakken, *Advan. Chem.* **33**, 86 (1961).
3. P. C. Aben, *J. Catal.* **10**, 224 (1968).
4. J. E. Benson, H. S. Hwang, and M. Boudart, *J. Catal.* **30**, 146 (1973).
5. A. Solbakken and L. H. Reyerson, *J. Phys. Chem.* **63**, 1622 (1959); **64**, 1903 (1960).
6. R. W. Zuehlke, M. Skibba, and C. Gottlieb, *J. Phys. Chem.* **72**, 1425 (1968).
7. L. H. Little, "Infrared Spectra of Adsorbed Species," p. 86. Academic Press, New York, 1966.
8. M. H. Dilke, D. D. Eley, and E. B. Maxted, *Nature (London)* **161**, 804 (1948).
9. T. J. Gray and C. C. McCain, *in* "Solid/Gas Interface II" (J. H. Schulman, ed.), p. 260. Butterworth, London, 1957.
10. W. C. Elmore, *Phys. Rev.* **54**, 1092 (1938).
11. R. Kaiser and G. Miskolczy, *J. Appl. Phys.* **41**, 1064 (1970).
12. T. Chirulescu and E. Segal, *Rev. Roum. Chim.* **13**, 1577 (1968).
13. A. E. Berkowitz and J. A. Lahut, *AIP Conf. Proc. Magnetism and Magnetic Materials* **10**, (2), 966 (1973).
14. L. Y. Sadler III and W. J. Hatcher, Jr., *J. Catal.* **38**, 73 (1975).
14a. U. Illgen and J. Scheve, *Z. Phys. Chem.* (L)**255**, 57 (1974).
15. P. Fouilloux, G.-A. Martin, A. J. Renouprez, B. Moraweck, B. Imelik, and M. Prettre, *J. Catal.* **25**, 212 (1972).
16. J. I. MacNab and R. B. Anderson, *J. Catal.* **29**, 328 (1973).
17. R. J. Kokes and P. H. Emmett, *J. Amer. Chem. Soc.* **81**, 5032 (1959).
18. J. I. MacNab and R. B. Anderson, *J. Catal.* **29**, 338 (1973).
19. G.-A. Martin and P. Fouilloux, *J. Catal.* **38**, 231 (1975).
20. I. Nicolau and R. B. Anderson, Int. Symp. on the Characterization of Adsorbed Species in Catalytic Reactions, Ottawa, 1974.
21. K. Saito and E. Fuma, *Nippon Kagaku Zasshi* **82**, 1324 (1961).

Appendix

Symbols Used More Than Once

C — Curie constant
H — Magnetic field strength
K' — Anisotropy constant
k — Boltzmann constant
L — Avogadro's constant
M — Magnetization
M_0 — M_s at $T = 0$ K
M_0' — M_0 after vapor adsorption
M_s — Saturation magnetization
M_{sp} — Spontaneous magnetization
M_r — Remanent magnetization
m — Magnetic moment
m_p — Moment of a particle
m(Fe) — Moment of iron
m_B — Bohr magneton
N_H — Number of H atoms
N_p — Number of particles
n — Moles of any entity
n_p — Moles of particles
n(H) — Moles of hydrogen atoms
n(Fe) — Moles of iron (atoms)
n(O$_2$) — Moles of oxygen molecules

S — Spin quantum number
T — Thermodynamic temperature
T_C — Curie temperature
T_N — Néel temperature
V — Volume
V_m — Molar volume
v — Volume of a particle
β — Bohr magneton number $= m/m_B$
β(Fe) — Bohr magneton number of iron
Δ — Weiss constant
ϵ — Change in β caused by adsorption
ϵ_{Ni}(H) — Change in β caused by adsorption of 1 mole H atoms per mole Ni
ζ — Bond number
ζ_{Ni}(C$_2$H$_6$S) — Bond number of dimethyl sulfide adsorbed on nickel $= \epsilon_{Ni}$(C$_2$H$_6$S)/ϵ_{Ni}(H)
κ — Magnetic susceptibility $= M/H$
ρ — Density
τ — Relaxation time

Definitions and Units in Magnetism

The four magnetic quantities used in this book are field strength (H), magnetization (M), moment (m), and susceptibility (κ).

There are several ways in which magnetic quantities may be defined. The method adopted is the familiar Gaussian-cgs. Another method is the Rational-mks. Definitions of the quantities are not the same in the two methods. Conversion factors for corresponding quantities are given below. To change a Gaussian-cgs quantity into a Rational-mks quantity replace the symbol at the left by the expression at the right:

$$H \qquad (4\pi\mu_0)^{1/2}H$$

$$M \qquad (\mu_0/4\pi)^{1/2}M$$

$$m \qquad (\mu_0/4\pi)^{1/2}m$$

$$\kappa \qquad \kappa/4\pi$$

The quantity μ_0 is the permeability of vacuum which is equal to $4\pi \times 10^{-7} \mathrm{J \cdot A^{-2}}$ (SI Rational-mks), or to 1.000 dimensionless (Gaussian-cgs).

Units often used, as adopted for this book, are

H	oersted	Oe
M	oersted	Oe
m	oersted $\times$ cm^3	Oe$\cdot$cm^3
κ	1	

The unit gauss has often been used for H and sometimes for M. The gauss and the oersted are identical. Some authors use $\mathrm{cm^{-1/2} \cdot g^{1/2} \cdot s^{-1}}$ for the unit of magnetization and some use "moment$\cdot$cm^{-3}." Many authors use erg$\cdot$Oe^{-1}, or erg$\cdot$gauss^{-1}, for the unit of moment.

Dimensions of the units used in this book are

H	Oe	$= \mathrm{cm^{-1/2} \cdot g^{1/2} \cdot s^{-1}}$
M	Oe	$= \mathrm{cm^{-1/2} \cdot g^{1/2} \cdot s^{-1}}$
m	Oe$\cdot$cm^3	$= \mathrm{cm^{5/2} \cdot g^{1/2} \cdot s^{-1}}$
κ	1	(dimensionless)

For use with the International System, SI units, the definitions recommended are the Rational-mks. These units are

H A·m^{-1} (ampere per meter)

M A·m^{-1}

m A·m^2

κ 1

Conversion of magnetic quantities from Gaussian-cgs to SI Rational-mks and *vice versa* involves a redefinition of the quantities themselves and a conversion of the units. Such correlations may be made as follows:

	Gaussian-cgs		SI Rational-mks
A field, H, of	1 Oe	is the same as	$(10^3/4\pi)$ A·m^{-1}
A magnetization, M, of	1 Oe	is the same as	10^3 A·m^{-1}
A moment, m, of	1 Oe·cm^3	is the same as	10^{-3} A·m^2
A susceptibility, κ, of	1	is the same as	4π

Author Index

D

Dalmai-Imelik, G., 49(18), 51(18), *53*, 57(3), 60(3), 61, *65*, 126(14), *135*
Dalmon, J.-A., 49(19), 51(19), *53*, 62(9), 63(13), 64(13), *65*, 104(1, 3), 105, 109, *117*, 129, *135*
Davis, B. J., 122(8), 126(15), *135*
de Boer, N. H., 59, *65*, 84(9), *94*, 115, *117*
Debye, P., 18, *29*
de Montgolfier, P., 49(19), 51(19), *52*, *53*, 60, *65*, 78(9), *79*, 93(14), *94*, 104(2), *117*
de Mourgues, L., 49(15), 51(15), *52*
Demuth, J. E., 126, 129, *135*, 143, *148*
Den Besten, I. E., 81(2), 87, *94*, 107(8), 110, 111, 113, *117*, 138, 140(4), *148*
Déportes, J., 63, 64, *65*
Deuss, H., 100, *102*
Dietz, R. E., 23, *29*, 32, 36, 38, *44*, 46(6), 48, 49, 50, *52*, 56, 57, *65*, 70(3), 72, 75, 77(3), 78(3), *79*
Dilke, M. H., 1, *15*, 151, *160*
Dorfman, J., 18, *29*
Dorgelo, G. J. H., 84(9), *94*
Dreyer, H., 48(11), 49, *52*
Dubinin, V. N., 27(30), *29*, 50(22), *53*
Dumesic, J. A., 64, *65*
Dutartre, R., 63(13), 64(13), *65*

E

Eastman, D. E., 126, 129, *135*, 143, *148*
Eggertsen, F. T., 42(15), 43, *44*
Eisehens, R. P., 98(4), *102*, 109, 111, 116, *117*, 119, 121, 123, 131, *135*
Eley, D. D., 1(2), *15*, 44(2), *44*, 123, *135*, 151, *160*
Elmore, W. C., 18, *29*, 154, *160*
Emmett, P. H., 115, *117*, 156, *160*
Erkelens, J., 131, *135*

F

Faessler, A., 99, *102*
Faraday, M., 40(7), *44*
Figueras, F., 49(15), 51(15), *52*
Forrer, R., 32(1), 38, *44*

Fouilloux, P., 33(4), *44*, 49(20), 51(20), *53*, 155, 156, 157, 158, *160*
Fox, P. G., 107(8), 110(8), *117*
Frackiewicz, A., 98, *102*
Francis, S. A., 109(10), *117*
Franzen, P., 23, *29*
Freel, J., 122, *135*
Frenkel, J., 18, *29*
Fulde, P., 24(15), *29*
Fuma, E., *160*

G

Galwey, A. K., 122, *135*, 138, *148*
Gans, R., 18, *29*
Garland, C. W., 109, *117*
Garnett, J. L., 144, *148*
Geus, J. W., 70, *79*, 83, 84, *94*, 106, 108, *117*
Gittleman, J. I., 64(16), *65*
Gomer, R., 100, *102*
Gottlieb, C., 151(6), 152, *160*
Gray, T. J., 153, 154, *160*
Grimley, T. B., 100, *102*
Guarnieri, C. R., 25(18), *29*

H

Halsey, G. D., Jr., 116, *117*
Hanak, J. J., 64(16), *65*
Hatcher, W. J., Jr., 155(14), *160*
Hayward, D. O., 84(7), *94*, 98, *102*
Helms, C. R., 62(8), 63(8), *65*
Henning, W., 26, *29*
Heukelom, W., 48, *52*, 67, *79*
Hill, F. N., 42(13), *44*
Hirota, K., 111, *117*
Hobson, J. P., 4, *15*
Holm, V. C. F., 42(14), *44*
Horiuti, J., 100, *102*, 130, *135*
Hwang, H. S., 150(4), *160*

I

Illenberger, A., 26, *29*
Illgen, U., 155, *160*
Imelik, B., 42(16), 43(16), *44*, 49(18, 20), 51(18, 20), *53*, 56, 57(3), 58, 60(6), 61, 62(7, 9), 63(7, 12), *65*, 78(9), *79*, 93(14),

Subject Index

A 5
B 6
C 7
D 8
E 9
F 0
G 1
H 2
I 3
J 4